T0201527

Analysis of Chemical Warfare Degradation Products

Analysis of Chemical Warfare Degradation Products

Karolin K. Kroening
Renee N. Easter
Douglas D. Richardson
Stuart A. Willison
Joseph A. Caruso

A John Wiley & Sons, Ltd., Publication

Registered office
John Wiley & Sons Ltd, The Atrium, Southern Gate, Chichester, West Sussex, PO19 8SQ, United Kingdom

For details of our global editorial offices, for customer services and for information about how to apply for permission to reuse the copyright material in this book please see our website at www.wiley.com.

Library of Congress Cataloging-in-Publication Data
Kroening, Karolin K., 1974- author.
  Analysis of Chemical Warfare Degradation Products / Karolin K. Kroening,
Renee N. Easter, Douglas D. Richardson, Stuart A. Willison, Joseph A. Caruso.
    p. cm
  Includes bibliographical references and index.
  ISBN 978-0-470-74587-8 (hardback)
  1. Organophosphorus compounds–Deterioration. 2. Chemical agents (Munitions)–
Deterioration. 3. Chemical agents (Munitions)–Analysis. 4. Decomposition
(Chemistry) I. Title.
  UG447.5.O74K76 2011
  623.4′592–dc22

                                                                              2010053414

A catalogue record for this book is available from the British Library.

Print ISBN: 9780470745878
e-PDF ISBN: 9781119993698
o-Book ISBN: 9781119993681
e-Pub ISBN: 9780470745878
Mobi ISBN: 9781119993698

Set in 12.5/15 Sabon by Laserwords Private Limited, Chennai, India.
Printed and bound in the UK by TJ International Ltd, Padstow, Cornwall.

About the Authors

Karolin K. Kroening, Ph.D.
University of Cincinnati, Cincinnati, OH, USA

Karolin received her Master's degree in 2006 from the University of Bologna, Italy, for research based on hydroxyapatite/chitosan composites for bone substitution. At the University of Cincinnati her research focused on the identification and cytotoxicity of chemical warfare agent degradation products and protein phosphorylation studies on cerebral spinal fluid, a study that may help in the development of drugs for patients diseased with a hemorrhagic stroke. She obtained her Ph.D. in December 2010 and is currently working for Procter and Gamble in Cincinnati, OH.

Renee N. Easter
University of Cincinnati, Cincinnati, OH, USA

Renee Easter earned a B.S. from Xavier University, Cincinnati, in 2007 and is currently pursuing a Ph.D. in Analytical Chemistry from the University of Cincinnati. Her research has focused on metallomics approaches to identifying proteins associated with cerebral vasospasm, as well as using internal tags, such as sulfur and phosphorus for identification and quantification of oligonucleotides for siRNA drug applications.

Douglas D. Richardson, Ph.D.
Merck Research Labs, Rahway, NJ, USA

Doug earned his B.S. in Forensic Chemistry with a minor in Biological Sciences from Ohio University in 2003. Following graduation Doug pursued his Ph.D. in the laboratory of Joseph A. Caruso at The University of Cincinnati. His research centered around advancements in elemental speciation, coupling a variety of separation techniques with element specific detection. This research was the first to utilize chromatographic techniques with inductively coupled plasma mass spectrometry for the analysis of nerve agent degradation products. In 2007, Doug defended his dissertation, earning his Ph.D. in Analytical Chemistry. Doug currently supports the development of novel pharmaceuticals within Merck Research Labs.

Stuart Willison, Ph.D.
National Homeland Security Research Center at the US Environmental Protection Agency, Cincinnati, OH, USA

Stuart Willison received his Ph.D. in Chemistry from the University of Cincinnati. He is currently working for the National Homeland Security Research Center at the US Environmental Protection Agency in Cincinnati, OH. His work involves environmental restoration following homeland security events, such as providing support in the detection, response to, and remediation of an area from a terrorist attack or an environmental disaster. Research areas include water protection and indoor/outdoor decontamination as well as method development of chemical warfare agent degradation products in various environmental matrices.

Joseph A. Caruso, Professor
University of Cincinnati, Cincinnati, OH, USA

Joe Caruso holds a Ph.D. from Michigan State University. After a one-year postdoctoral fellowship at The University of Texas – Austin, he joined the University of Cincinnati Chemistry faculty and since then he has authored or co-authored 380 scientific publications and presented more than 325 invited lectures at universities, scientific meetings, government and industry laboratories. His current research interests are in: metallomics studies involving transgenic plants and their phytoremediation mechanisms or enhancements; evaluating cell signaling changes through phospho- or metallo-proteomes as biomarkers in the CSF of certain stroke patients; investigating the metalloproteomes associated with viruses and their effect on viral capsid stability; and the effects on cell signaling changes when arsenic toxified cells are given selenium species as part of the nutrient mix.

Joe Caruso is a member of the American Chemical Society, Society for Applied Spectroscopy and a Fellow of the Royal Society of Chemistry (RSC). He is Chair of the RSC Metallomics Editorial Board. He has been honored many times including the 2000 Spectrochemical Analysis Award given by the Analytical Division of the American Chemical Society, the University of Cincinnati – Excellence in Doctoral Student Mentoring Award in 2006, and in 2007 he received the Rieveschl Award for Distinguished Scientific Research. His most recent award was to be elected Fellow of the Society of Applied Spectroscopy.

Contents

Preface

Lethal chemical warfare agents, including nerve agents and vesicants, still pose major threats to life around the world and our surrounding environment. Though their use has been forbidden by international conventions, nerve agents and vesicants are still produced and stockpiled by terrorist organizations. These agents degrade relatively easily. Therefore, it is understandably of great importance that these agents of interest and their degradation products are detected. Rapid and sensitive methods are required, in order to identify these warfare agents and their degradation products. This book describes the chemistry of nerve agents and vesicants, their decomposition and degradation products, in addition to their toxicity, and includes a list of appropriate detection and analysis techniques. Also included is a brief history of the research area, separation techniques, detection methods and detection limits together in a short, easy to read text, with an adequate number of tables and references for the reader who is looking for further detail.

The work to prepare this book was undertaken by the Caruso research group at the University of Cincinnati, including current and former graduate students, who, through their graduate studies, amassed a high degree of knowledge regarding warfare agents and the analysis of their degradation products through various analytical techniques. While we do not pretend to portray all analytical techniques and methods currently in use, for some may be proprietary or classified, our hope is that those who are generally interested in warfare agents will profit from this text. This includes those wishing to learn about analysis, also environmentalists, and more generally, those who have interest in small molecule phosphorus, sulfur and arsenic chemistry beyond warfare agents and their decomposition products, as in pesticides or herbicides. Overall, we have aspired to produce a product that will be of practical use as well as a motivating factor for continued research interest in this field.

Renee N. Easter
Karolin K. Kroening

1

Historical Milieu

Analysis of Chemical Warfare Degradation Products, First Edition. Karolin K. Kroening,
Renee N. Easter, Douglas D. Richardson, Stuart A. Willison and Joseph A. Caruso.
© 2011 John Wiley & Sons, Ltd. Published 2011 by John Wiley & Sons, Ltd.

1.1 ORGANOPHOSPHORUS NERVE AGENTS

Organophosphorus (OP) type compounds, that is, derivatives containing the P=O moiety, were first discovered in the 1800s when researchers were investigating useful applications for insecticides/rodenticides. There are many derivatives of organophosphorus compounds, however, the OP derivatives that are typically known as 'nerve agents' were discovered accidentally in Germany in 1936 by a research team led by Dr. Gerhard Schrader at IG Farben [1–4]. Schrader had noticed the effects and lethality of these organophosphorus compounds towards insects and began developing a new class of insecticides. While working towards the goal of an improved insecticide, Schrader experimented with numerous phosphorus-containing compounds, leading to the discovery of the first nerve agent, Tabun (or GA) (Figure 1.1).

The potency of these insecticides towards humans was not realized until there was yet another accident, which involved a Tabun spill. Schrader and coworkers began experiencing symptoms, such as miosis (constriction of the pupils of the eyes), dizziness and severe shortness of breath, with numerous effects lasting several weeks [1, 4, 5].

A number of years prior to Schrader's discovery, Adolf Hitler and the Nazi government required that all inventions with potential military significance must be reported to the government [1, 3, 4]. After Schrader's discovery, the government secretly became involved in the development of chemical weapons for military use and built several large classified facilities during World War II for the further investigation

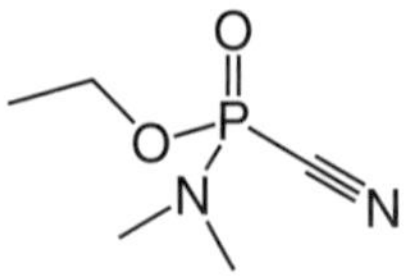

Figure 1.1   Chemical structure of Tabun

of these chemicals and the possibilities of their effectiveness as chemical weapons. Concentrated focus on this research area led to the discovery of other nerve agents, such as Sarin, which was discovered by Schrader and his team in 1938. It is believed that SARIN was creatively named using the initials of these workers: Schrader, Ambrose, Rudriger and van der Linde. Soman [4] and Cylcosarin were also discovered through the German research efforts during World War II.

The Allies did not learn of nerve agents until artillery shells filled with them were captured towards the end of the war. After the Allies seized control of various chemical plants and uncovered some of Germany's chemical weapons, they too began to develop research on nerve agent chemical weaponry [1]. The G-series naming system was created by the United States when it uncovered the German activities, labeling Tabun as GA (German Agent A), Sarin as GB, Soman as GD and Cyclosarin as GF. Studies suggest that further derivatives of the G-agents (derivatives of Sarin) were made either by Germany or by the Allies after they had begun research on chemical weapons [1, 5]. Because most of the lesser known agents were not mass produced, or compounded, and by the fact that most of the work was done in secrecy, there is relatively little literature on the lesser known Sarin derivatives. The German government incorporated several of these nerve agents into artillery shells, but never used them against any Allied targets. It was believed that German intelligence was aware that the Allies also possessed the knowledge of chemical warfare agents, or similar compounds, which deterred the Germans from using the agents for fear of extensive retaliation, which would perhaps have resulted in their own nerve agents being used against them [2, 5].

After the end of World War II, the Allied nations began to divide the captured chemical weapons amongst themselves, and continue research on chemical nerve agents [5].

Dr. Ranajit Ghosh was a scientist in the United Kingdom working as a chemist at Imperial Chemical Industries (ICI). Similarly to the organophosphate research performed by Schrader, Ghosh also found the compounds to be relatively effective pesticides. Ranajit Ghosh and J.F. Newman discovered a new organophosphorus compound at ICI in 1952, patenting the first V-agent, diethyl S-2-diethylaminoethyl phosphorothioate (VG). At the time, its potency was not fully understood and it was subsequently sold as a pesticide under the trade name Amiton. It was withdrawn a short while later, due to the fact that it was found to be too toxic for safe use.

As with Schrader's work with the first G-agent, the toxicity did not go unnoticed, and samples were sent to the British Armed Forces research facility in England for extensive evaluation. Only after further experiments was the agent Amiton renamed to VG. Through British research efforts, a new series of organophosphorus compounds were discovered, which were known as the V-series [5]. However, once the aggressive lethality of these compounds towards humans was realized, the British government halted all chemical weapons research within a few years. The V-series agents were considered to be some of the deadliest manmade substances, where only a single drop is enough to kill an adult. The research was not abandoned though, as the United States and the British governments decided to exchange information regarding the V-series technology for research on thermonuclear weapons [2–5]. The United States obtained knowledge of four agents, VG, VE, VM, and the most popular V-agent, VX, and began experimenting while producing large stockpiles of weaponized agents [5]. In addition to being known as V-agents, they are occasionally referred to as Tammelin's esters, after Lars-Erik Tammelin of the Swedish Institute of Defense Research, who was also conducting research on this class of compounds in 1952 [2, 4, 5]. He is not often cited for this research, mainly because his work was not widely publicized.

The United States was not the only other country interested in the investigation of organophosphorus nerve agents. Russia, too, had developed its own V-agent, known as Russian VX, or VR [5]. Not much is known about these agents in terms of research due to the secrecy of the former Soviet Union. Russian VX has similar properties to those of VX and is similar in toxicity as well. Other agents, known as Novichok agents, were also produced, although again not much is known about these agents either. It is presumed that the Novichok agents were initially more stable and less toxic versions to handle than the earlier nerve agents, simply because it was assumed that they worked as binary agents [5]. Once two binary agents are mixed, then the toxic and lethal agent is produced. As with the G-agents, much of the research involving V-series agents was carried out in secrecy. Other derivatives of the V-series agents are likely to exist, but probably demonstrate less toxic properties than VX or VR, or were simply not made in vast quantities. Thus, there is no accurate information of their existence.

1.2 BLISTER AGENTS

Blister agents are cytotoxic alkylating compounds and have the ability to produce chemical burns on the skin; they are chemicals that produce blisters. Blister agents are often also called vesicants, the etymology for which originates from Latin: *vesica* = bladder, blister [6].

Some examples of blister agents are: Lewisite, an organo-arsenic compound; Sulfur Mustard, also known as Mustard Gas, a family of sulfur-based agents; Nitrogen Mustard, a family of agents with similar characteristics to the Sulfur Mustards, but nitrogen based; and Phosgene Oxime, a potent chemical warfare agent, dichloroformoxime, which falls under the CDC (Center for Disease Control and Prevention) category of blister agents, and is often referred to as nettle agent. Nettle agents (or urticants) produce corrosive skin conditions, such as urticaria, but not cutaneous blisters, therefore their classification as blister agents is often debatable [7, 8].

Lewisite was discovered over one hundred years ago, in 1903. In his doctoral thesis, Father Julius Arthur Nieuwland, Ph.D. (14 February 1878 to 11 June 1936), working at the Catholic University of America in Washington, DC, described his study of the reaction of acetylene gas with arsenic trichlorides in the presence of aluminum chloride [9] (Figure 1.2).

During the course of this work, he discovered Lewisite and was himself exposed to it and became ill for several days. Therefore, he decided to abandon the study in order to recover from the illness caused by the toxic substance created during his experiments, and also due to its potential use as a toxin.

During World War I, the main American Chemical Warfare Service (CWS) unit, the American University Experiment Station (AUES), was conducting research and experiments in order to develop new chemical warfare agents. The information from Nieuwland's thesis became of interest to Winford

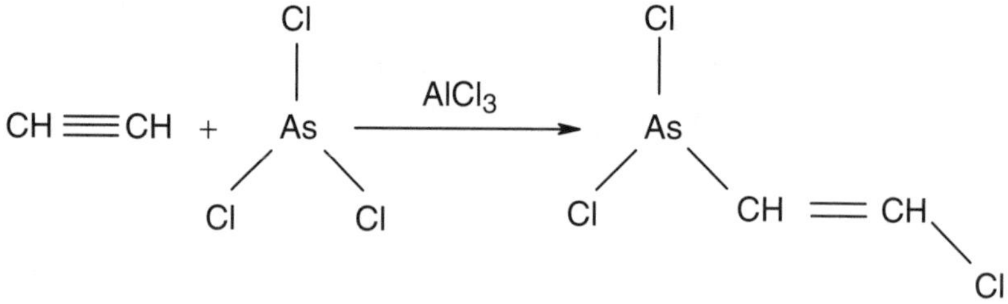

Figure 1.2   Reaction scheme of Lewisite

Lee Lewis, who was an associate professor of chemistry at Northwestern University before he volunteered for CWS research in Washington. He tried to repeat Nieuwland's experiments and, after some adjustments (such as the addition of hydrogen chloride before distillation), discovered that the mixture was composed of three similar arsenic-based compounds, which, depending on the number of acetylene molecules bonding with the arsenic trichloride, became known as: Lewisite 1, 2-chlorovinylarsonous dichloride; Lewisite 2, bis(2-chlorovinyl)arsinous chloride; and Lewisite 3, tris(2-chlorovinyl)arsine [5, 10, 11]. The CWS named the arsenic compounds after Lewis (1917), which turned out to be one of the most deadly poisonous gasses produced during World War I. It was also known as the 'Dew of Death', because it was often disseminated by air, and as shown on the poster in Figure 1.3, because of its strong smell that resembled geraniums.

Although the USA produced Lewisite during World War I, it was also discovered independently in Germany, in early

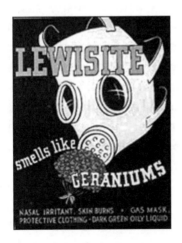

Figure 1.3 Identification poster for Lewisite (1941–1945). Picture kindly provided from the National Museum of Health and Medicine, Washington D.C., USA

1918, but it was rejected in favor of ethyldichloroarsine [12]. During World War II, the United States, Great Britain, the Soviet Union, Germany and Japan had large supplies of Lewisite available [13]. Following World War II, 2,3-dimercaptopropanol, (INN: dimercaprol or 'British anti-Lewisite') was developed by British biochemists as an antidote to Lewisite. Thus, this agent was then considered to be outdated by the major war powers and was no longer considered for weapons purposes [14, 15]. Today, dimercaprol is used medically in the treatment of arsenic, mercury, gold and lead, and other toxic metals, poisoning, as it is a heavy metal chelator [16]. Although it is generally considered that Lewisite has not been used since World War I, it may have been used, in addition to a mustard agent, more recently by Saddam Hussein against the Iraqis [17–19].

Sulfur Mustard is a family of sulfur-based chemical warfare compounds. Sulfur Mustard, 1,5-dichloro-3-thiapentane, is the most well known due to its frequent use, especially during World War I, and has numerous synonyms, such as HD, Senfgas, Sulfur Mustard, blister gas, s-lost, lost, Kampfstoff LOST, yellow cross liquid and Yperite [20]. In 1822, Despretz reacted ethylene with sulfur chloride to yield bis(2-chloroethyl) sulfide. Over the next 100 years, at least six other reactions for the production of Sulfur Mustard were devised. Each yielded Sulfur Mustard but differed in the immediate precursors, the purity of the mustard produced, and the specific byproducts. Sulfur Mustard was used for the first time by the German forces in World War I during the battle near Ypres, Belgium, to combat the Canadian army (1917). Sulfur Mustard was also used by the Japanese against the Chinese early in World War II. During the 1980s, Iraq used these CWAs (chemical warfare agents) twice: first, in its war with Iran, and second, against its Kurdish minority in Northern Iraq. As a consequence, American soldiers were exposed to Iraqi chemical agents when arms dumps were destroyed during and after the 1991 Gulf War. A timeline and brief history of Sulfur Mustard can be found in Table 1.1 [21].

After World War I and the deadly use of Sulfur Mustard, its production appeared to increase, and the methods of synthesis somehow enhanced its toxicity. During the period 1917–1945 over 200 000 tons of 'mustard' were produced

Table 1.1 Timeline and history of Sulfur Mustard discovery and use [21]. Table reproduced with the permission of Prehospital and Disaster Medicine, Madison, WI; An Official Publication of the World Association for Disaster and Emergency Medicine

1822	Belgian scientist Cesar-Mansuete Despretz synthesized an impure uncategorized form of Sulfur Mustard.
1886	German chemist Victor Meyer created the Levinstein process to purify Sulfur Mustard. This distilled pure form is known as HD.
1917	The Germans utilized Sulfur Mustard for the first time as a chemical warfare (CWA) agent at Ypres, Belgium. The type of Sulfur Mustard used during World War I was called Hun Stoffe, abbreviated to HS or H, and contained 20–30% impurity. During this period, soldiers were equipped only with respiratory protection devices. Inadequate skin protection resulted in over 1.3 million people receiving Sulfur Mustard related injuries during World War I with >90 000 dying.
1935–1936	Italy conquers Ethiopia using aircraft delivery of Sulfur Mustard.
1937–1945	Japan invades China using chemical warfare agents that include Sulfur Mustard.
1963–1967	Egypt intervenes in the Yemen Civil War by using Sulfur Mustard aerial bombs against royalist forces.
1983–1988	Iraq uses Sulfur Mustard and nerve agents in the Iran–Iraq conflict.
1987–1988	Iraq uses Sulfur Mustard against Kurdish fighters.

by more than ten countries, including Canada, Japan, France, Poland, Germany, the United Kingdom, Hungary, the United States, Italy, the USSR and South Africa. Since World War II, at least another six countries are known to have produced mustards in one form or another. A cursory review of this production reveals that, in addition to Sulfur Mustard, at least six other primary compounds have been produced. These primary compounds included the following [22]:

- bis(2-cholorethyl)sulfide – basic Sulfur Mustard;
- 1,2-bis(2-chloroethylthio)ethene – Sesqui Mustard;
- bis(2-chloroethylthioethyl)ether – Oxygen Mustard;
- sulfur mustard with a methyl group – Ziakov Mustard;
- bis(2-chloroethyl)ethylamine – Nitrogen Mustard-1 (HN-1);
- methyl-bis-(2-chloroethyl)amine – Nitrogen Mustard-2 (HN-2);
- tris(2-chloroethyl)amine – Nitrogen Mustard-3(HN-3).

While Sulfur Mustard was used extensively in World War I and World War II, its relative Nitrogen Mustard was stockpiled by several nations during the Second World War, but never used in combat. All Nitrogen Mustards (HN-1, HN-2 and HN-3) were synthesized in the 1930s, but were not produced in amounts large enough for warfare. Mechlorethamine (HN-2: Mustargen) became a prototypical cancer chemotherapeutic compound and remained the standard compound for this purpose for many years [23].

There is not as much information or historical data on Phosgene Oxime, CX, (not to be confused with phosgene gas) as there is for Sulfur Mustard and other blister agents. It was first synthesized by the Germans in 1929. Its name appears frequently in military chemical literature, but there is no current assessment of the potential of Phosgene Oxime as a military threat agent [18]. CX is of military interest because it penetrates garments and rubber much more quickly than other chemical agents, and it produces the rapid onset of severe and prolonged effects.

1.3 STERNUTATOR AGENTS

Chemical warfare agents have been used since the beginnings of the 1900's. One of the earliest reports of irritant use is from the Russo–Japanese war (1904–1905), when the Japanese forces used arsenical rag torches made from long bamboo to form a choking cloud over the Russian trenches [24]. In the Siege of Petersburg, Virginia, during the American Civil War, a suggestion was made to President Lincoln to use a mixture of sulfuric acid and muriatic acid, which would form a dense cloud covering the Confederate troops. This mixture would also render the opposing forces incapacitated by sneezing, coughing and tearing [24]. However, it was not until World War I, that the sternutator agents known today as Clark I, Clark II and Adamsite were introduced into warfare.

Diphenylchloroarsine (DA, Clark I) was first synthesized in 1881 by Michaelis and Lacoste. These researchers described the exposure effects of powerful irritations of the nose and eyes as well as blistering of the skin. The vapor caused respiratory complications, fainting and paralysis in the extremities [25]. Sturniolo and Bellinzoni developed diphenylcyanoarsine (DC, Clark II) in 1918. Similar in structure to Clark I, it also has the same exposure effects. Wieland, a German scientist, was the first to synthesize 10-chloro-5,10-dihydrophenarsazine (DM) in 1915. Three years later a team in the USA independently developed the same chemical and named it Adamsite, after Major Roger Adams, the team leader [26].

Although under moderate exposure these agents are not toxic, they played a key role in World War I. All three of the agents (Clark I, Clark II and Adamsite) are solid at room temperature and when incinerated and condensed, a fine aerosol is created that could pass through the protective filters of the gas masks of that era [26]. The term 'mask breakers' came from this unique property and when these agents were combined with more toxic nerve agents, the combination was deadly. The sneezing, coughing and vomiting would force soldiers to expel their protective gear rendering them helpless against other toxic gases. Glass bottles containing the agents (Figure 1.4) were often enclosed in shells that could be launched, causing the bottle to break at the target and the agent to be released.

The technique of 'mask breaking' by diphenylchloroarsine (DA) was first introduced in July of 1917 by German forces, while diphenylcyanoarsine (DC) was later used in May 1918.

Figure 1.4   A glass bottle containing Clark I from World War II. Typically, the bottle would be enclosed in a shell casing and when it exploded the glass would break, vaporizing the liquid, causing its release. Picture kindly provided by The Museum of Technology, Hemel Hempstead, Hertfordshire, UK

These agents, along with Adamsite, were typically used on civilians and as riot control agents up until the 1930s, when they were banned by Western nations due to the toxicity of the degradation products [27]. Some nations still continued to use these agents even after the ban. Japan used sternutator agents in their invasion of China from 1937 to 1945 and South Vietnam forces employed Adamsite during the Indo–China intervention in the 1960s [26]. The use of these agents eventually declined steadily, and by 1967 the technology of the protective masks had advanced to the point that the aerosol was filtered out, protecting the soldier from exposed to exposure [27]. Although less toxic than other warfare agents, sternutators have played a major role in modern warfare.

1.4 CHEMICAL WEAPONS CONVENTION (CWC)

As a consequence of prior events, many countries possessed nerve agents, which led to the development of the Convention on the Prohibition of the Development, Production, Stockpiling and Use of Chemical Weapons and on their Destruction (more commonly known as the Chemical Weapons Convention, CWC). The CWC evolved into the development of a treaty covering 188 countries, including the United States. After the formation of the CWC, an arms control agreement was drafted in 1992, signed on 13 January 1993, and became effective as international law on 27 April 1997. It prohibits the stockpiling, production and use of chemical weapons; however, it does not outlaw the production or use of related chemicals for other purposes. The Organization for the Prohibition of Chemical Weapons (OPCW) is the current administrator of the treaty, and is also in charge of the regulatory inspections in each member country that are set forth in the treaty. Aside from inspections, the treaty also breaks down three classes of chemicals, based on the toxicity of the chemical directly or for the purpose of being used as a chemical weapon.

1.4.1 Schedule of Chemicals

Schedule 1 chemicals are toxic chemicals or precursors that have been used, stockpiled or developed as chemical warfare agents. Examples of these are Sarin, VX and Sulfur Mustard. Also in this category is any chemical precursor that has a related structure, related properties, including toxicity, to a chemical found in Schedule 1 and that has very little usefulness for purposes not prohibited by the CWC.

Schedule 2 compounds are chemicals that pose a threat or risk to the CWC due to intrinsic toxicity. These chemicals could be used as possible chemical weapons or Schedule 1 chemicals. Examples of Schedule 2 chemicals are Amiton, Perfluoroisobutene (PFIB) and 3-quinuclidinyl benzilate (BZ). These chemicals are typically not mass produced for any purpose.

Chemicals found in Schedule 3 are known to be toxic and pose a risk because of their importance in the production of Schedules 1 or 2. These chemicals, unlike Schedule 2, are often produced commercially in large amounts for peaceful purposes. Schedule 3 chemicals include phosgene, cyanogen chloride and chloropicrin. The agents discussed in this book and their classes can be seen in Table 1.2.

1.4.2 Destruction of Chemical Weapons

As part of the treaty, a member state must declare any stockpiles or stored chemical weapons. The CWC guidelines call for each member to report every storage facility by name and geographical location, in addition to a detailed inventory

Table 1.2 CWC schedule of CWA

Agent	Appearance	CWC Schedule
Tabun	Clear, colorless liquid, faint fruity smell	1
Soman	Clear, colorless liquid, slight camphor odor	1
Sarin	Colorless, odorless liquid	1
VG	Amber colored oily liquid	2
VE	Amber colored oily liquid	1
VM	Amber colored oily liquid	–
VX	Amber colored oily liquid	1
Lewisite	Oily, colorless liquid when pure	1
Sulfur Mustard	Can be clear yellow or brown as liquid	1
Nitrogen Mustard	Can be clear or pale amber in color with fishy, musty, or fruity smell	1
Phosgene Oxime	Colorless solid, yellow–brown liquid	–
DA	White solid	–
DC	Colorless crystals, garlic smell	–
DM	Yellow to green crystals	–

list and site diagram for all facilities. Each member state is responsible for the destruction of their chemical weapons, as described in the guidelines of the CWC. Nations are not allowed to dump chemical weapons in a body of water, bury them in land, or burn the chemical weapons as a means of destruction. The CWC is in place in order to protect people and the environment from chemical weapons, their degradants and to provide guidelines for safely removing the threat.

Since 1997, all nations that are part of the CWC have agreed to destroy all stockpiles of chemical nerve agents.

Thus far, only Albania has completely destroyed 100% of its stockpile, being the first nation to do so.

Because every nation has not signed the treaty, concern still remains with respect to the countries unwilling to join and to terrorists groups who may have been able to acquire such agents through one of these nations. There have been instances when terrorists groups or countries have used nerve agents. In 1995, a terrorist group, known as the Aum Shin-rikyo religious group, released Sarin into the Tokyo subway. During the Iran–Iraq war (1981–1988), it is suspected that nerve agents were used against Iran. These examples indicate how important the treaty is and why the nations that are a part of the treaty are willing to destroy such harmful and dangerous chemicals.

REFERENCES

[1] Borkin, J. (1978) *The Crime and Punishment of IG Farben*, Free Press, New York, and references therein.
[2] Black, R.M. and Harrison, J.M. (1996), in *The Chemistry of Organophosphorus Compounds*, vol. 4 (ed. F.R. Hartley), John Wiley & Sons Inc., New York, pp 781–840, and references therein.
[3] Harris, R. and Paxman, J. (1982) *A Higher Form of Killling*, Hill and Wang, New York, and references therein.
[4] Robinson, J.P. (1971), in *The Problem of Chemical and Biological Warfare*, vol. 1, Humanities Press, New York, and references therein.
[5] Tucker, J.B. (2006) *War of Nerves: Chemical Warfare from World War I to Al-Qaeda*; Pantheon Books, New York, and references therein.
[6] McManus, J., Huebner, K. (2005) *Crit. Care Clin.*, 21, 707–718.
[7] Wilhelm, P., Werner, D. (1932) *Ber. Dtsch. Chem. Ges. (A and B Series)*, 65, 754–759.
[8] Department of Health and Human Services (2007) *Medical Management Guidelines for Phosgene Oxime*, Agency for Toxic Substances and Disease Registry.

[9] Paxson, F.L. (2005) *The Last American Frontier*, Kessinger Publishing.

[10] Vilensky, J.A., with Sinish, P.R. (2005) *Dew of Death: The Story of Lewisite, America's World War I Weapon of Mass Destruction*, Indiana University Press, Bloomington, IN.

[11] Vilensky, J.A., Redman, K. (2003) *Ann. Emergency Med.*, **41** (3), DOI: 10.1067/mem.2003.72.

[12] Epstein, J.R., Rosenblatt, D.H., Gallacio, A., McTeague, W.F. (1973) Summary report on a data base for predicting consequences of chemical disposal operations, *EASP 1200-12, January 1973, AD-B955399* (distribution limited to U.S. Government).

[13] Compton, J.A.F. (1988) *Military Chemical and Biological Agents: Chemical and Toxicological Properties*, Telford Press, Caldwell, NJ, pp. 37–38.

[14] Goldman, M., Dacre, J.C. (1989) *Rev. Environ. Contam. Toxicol.*, **110**, 75–115.

[15] Peters, R.A.S., Stocken, L.A., Thompson, R.H.S. (1945) *Nature*, **156**, 616.

[16] Mückter, H, Liebl, B., Reichl, F.X. *et al.* (1997) *Hum. Exp. Toxicol.*, **16**, 460–465.

[17] http://www.psgcabo.com/hottop.html#mustard (access date 3 December, 2010).

[18] Croddy, E., Wirtz, J., Larsen, J. (eds) (2005) *Weapons of Mass Destruction: An Encyclopedia of Worldwide Policy, Technology, and History*, ABC-CLIO, Santa Barbara.

[19] HistoryNet Staff MHQ (2006) Weaponry: Lewisite – America's World War I Chemical Weapon, *Quart. J. Military History*, HistoryNet.com (access date 20 June 20th, 2010).

[20] Mesilaakso, M. (2005) *Chemical Weapons Convention Chemical Analysis: Sample Collection, Preparation and Analytical Methods*, John Wiley & Sons, Chichester.

[21] Wattana, M., Tareg, B. (2009) *Prehosp. Disaster Med.*, **24**, 19–29.

[22] McGeorge, H.J. (1998) Mustard: An Ambiguous Term, Public Safety Group, May 28, 1998, http://www.psgcabo.com/hottop .html (accessed 3 December, 2010).

[23] Eric, C.V. (1984) *Int. J. Dermatol.*, **23**, 180–186.

[24] Romano, J., Jr., Lukey, B.J., Salem, H. (eds) (2008) *Chemical Warfare Agents, Chemistry, Pharmacology, Toxicology and Therapeutices*, 2nd edn, CRC Press, Taylor and Francis Group.

[25] NTI diphenylchloroarsine, http://www.nti.org/e_research/profiles/
nk/chemical/1094.html (accessed 24 February 2010).

[26] Szinicz, L. (2005) *Toxicology*, **214**, 167–181.

[27] Noblis (2010) A short history of chemical warfare during
World War I, http://www.noblis.org/MissionAreas/nsi/Background
onChemicalWarfare/HistoryofChemicalWarfare/Pages/History
ChemWarfareWWI.aspx (accessed 24 February 2010).

2

Toxicity of Chemical Warfare Agents and their Degradation Products

Analysis of Chemical Warfare Degradation Products, First Edition. Karolin K. Kroening,
Renee N. Easter, Douglas D. Richardson, Stuart A. Willison and Joseph A. Caruso.
© 2011 John Wiley & Sons, Ltd. Published 2011 by John Wiley & Sons, Ltd.

2.1 ORGANOPHOSPHORUS NERVE AGENT TOXICITY

The toxicity of organophosphorus nerve agents has been well documented [1–3]. Following the discovery of the synthetic procedure in the 1850s, the Germans were the first to produce organophosphorus nerve agents on a large scale, in the late 1930s [4]. Considered the most toxic of all chemical warfare agents (CWAs), the first reported use of organophosphorus nerve agents [Tabun (GA) and Sarin (GB)] was by Iraq against Iran during the first Gulf War and also against Kurdish rebels [4]. Sarin (GB) was also released during the Japan subway attack (1995) resulting in 12 deaths [4]. This chapter and the Appendix provide a detailed description of the toxicity mechanism, exposure route, lethal dose limits and medical treatment procedures for organophosphorus nerve agents and exposure to these toxic chemicals. A summary of nerve agent toxicity values is provided in Table 2.1.

2.1.1 Toxicity Mechanism – Acetylcholinesterase Inhibition

Organophosphorus nerve agents are potent cholinesterase inhibitors, specifically acetylcholinesterase (AChE), the

Table 2.1 Toxicity values for organophosphorus nerve agents. Table taken from Reference [5] with permission from *Environmental Health Perspectives*

Compound	LD$_{50}$ inhalation (mg per person)	LCt$_{50}$ skin (mg-min m^{-3})	CAS number
Tabun (GA)	1000	200	77-81-6
Sarin (GB)	1700	100	107-44-8
Cyclosarin (GF)	30	Unknown	329-99-7
Soman (GD)	50	70	96-64-0
VX	10	50	50782-69-9

enzyme responsible for central and peripheral nervous system communication with organs [2, 6]. Nerve agents react with the serine hydroxyl group on the active site of AChE resulting in the formation of phosphates and/or phosphonate esters [6]. The phosphorylation of the AChE enzyme results in decreased regeneration and inhibition of acetylcholinesterase at the cholinergic synapse junctions. This decrease results in excess accumulation of the neurotransmitter acetylcholine, leading to continuous stimulation of muscles (cholinergic crisis) and potential death by asphyxiation. Toxicity of nerve agents is directly related to the chirality of the phosphorus atom [7]. This basic mechanism of toxicity is common for all G-type nerve agents, which follow a hydrolysis metabolic pathway associated with A-cholinesterases [7]. While V-type agents are less prone to hydrolysis (a lower affinity for AChE), both types of agents can interact with additional serum cholinesterases and carboxylesterases in addition to proteins [6, 7]. V-type agents also produce additional organ toxicities due to their intrinsic physical properties.

2.1.2 Exposure

Organophosphorus nerve agents are liquids at room temperature [4]. This chemical property is often confusing, as these compounds are often referred to as 'nerve gases' or 'poison gases' [2, 3, 8]. A source of this misperception is associated with the likely exposure route in a warfare setting, inhalation due to aerosol exposure [9]. Inhalation is the most toxic form of exposure. Symptoms, including death, can appear within minutes following inhalation [10]. Alternative exposure routes, typically from liquid contact, include dermal, ocular and ingestion; however, these mechanisms are not as acute as aerosol inhalation, with the exception of an immediate ocular response to nerve agent vapor [4, 10]. Symptoms following liquid exposure can develop within 30 minutes up to 2 days after contact with the nerve agent. Typical symptoms associated with nerve agent exposure include papillary muscle contraction, gland secretion, bronchospasms, abdominal cramping, nausea, vomiting, diarrhea, heart rate changes, hypertension, seizures, hypoxia and death [2, 6]. Exposure prevention to inhalation and liquid contact include respirator and protective barriers such as a HAZMAT suit, as shown in Figure 2.1 [11].

2.1.3 Response, Treatment and Prevention

The initial response to nerve agent exposure consists of immediate decontamination, respiratory aid and targeted medicinal antidotes [4, 10]. Removal of clothing/protective layers and decontamination with water and sodium hypochlorite assists in the removal of residual vapors that may be trapped. Thorough decontamination of the skin with water and sodium hypochlorite is imperative. A US Food and Drug Administration (FDA) approved skin decontamination kit containing a charcoal impregnated ion-exchange resin is available [2].

Medical response, treatment and prevention (prophylaxis) to nerve agent exposure are ever expanding fields of research [6, 12, 13]. Atropine, which rapidly reverses cholinergic over expression at synapse junctions, is the most commonly used anticholinergic. Mark I kits, which contain 2 mg atropine with an autoinjector for intramuscular use, are issued to US military personnel [2]. Recommended initial field doses for atropine are 2, 4 or 6 mg with re-treatment every 5–10 min depending on the symptoms. In addition, oximes reactivate cholinesterase enzymes, ensuring normal function. Mark I kits also include 600 mg autoinjectors of 2-pralidoxime chloride (2-PAM Cl) with initial field doses of 600, 1200 or

(a) (b)

Figure 2.1 Powered Respirator Protective Suit (PRPS). Figure kindly provided by Respirex International Ltd, Redhill, Surrey RH1 4DP, UK

1800 mg [2]. Benzodiazepine is utilized to counteract seizures associated with nerve agent exposure. Typically administered in 10 mg autoinjector doses, midazolam or diazepam are approved for field use.

Prophylaxis (pre-treatment) is defined as the implementation of medical countermeasures prior to exposure of an organism to a nerve agent [6]. These countermeasures fall into two main categories: protection against AChE inhibition, and preventative action against acetylcholine accumulation (the anticholinergics described above). Sub-categories for protection against AChE inhibition include: scavengers, reversible AChE inhibitors and AChE reactivators [2, 6, 12].

The use of scavengers, typically proteins or antibodies, follows a detoxification pathway in which the nerve agent is bound or degraded prior to penetration of the target site (cholinesterase enzyme). Enzymes that bind nerve agents are typically referred to as stoichiometric scavengers, while enzymes that hydrolyze nerve agents are called catalytic scavengers [6]. A variety of cholinesterases have been utilized as scavengers against nerve agent exposure [12]. Reversible AChE inhibitors protect the normal function of AChE through a modification of the enzyme making it resistant to nerve agent exposure. These inhibitors, typically carbamates, are able to protect AChE against nerve agents and then restore enzyme function following decaramylation. Examples of reversible AChE inhibitors include Tacrine, Huperzine A, 7-MEOTA, Aminostigmine and Pyridostigmine (most common) [6]. Bajgar *et al.* [6] have provided a detailed description of pharmalogical prophylaxis against nerve agent exposure.

2.2 TOXICITY OF NERVE AGENT DEGRADATION PRODUCTS

Most of the degradation products for nerve agents are not as toxic as the parent agents themselves, but there are a few exceptions such as VX. All nerve agents are viscous liquids. The G-agents tend to be more volatile than their V-agent counterparts, presenting more of a vapor hazard, whereas V-agents tend to be more of a surface hazard. The G-agents are also more soluble in water than the V-agents and, therefore, more susceptible to hydrolysis.

2.2.1 Toxicity of GA (Tabun) Degradation Products

The nerve agent GA is simple to make, but is typically impure, requiring a purification process. In addition to the presence of impurities, GA could consist of numerous degradation products if it were to sit for a period of time, depending on the degradation pathways and environmental conditions. Of all of the G-agents, GA produces the greatest number of degradation products [5, 14–19]. Typically, GA usually consists of degradation products and/or other by products, and is rarely found in its pure form. Literature supports this statement as evidenced by D'Agostino and colleagues who analyzed munitions-grade samples of GA [16, 18]. They established that impurities accounted for approximately 28% of the volatile organic content, with the most prevalent impurity being diethyl dimethylphosphoramidate, particularly in soil samples. Other relative amounts of impurities found in GA are ethyl dimethylphosphoramidate, O-ethyl O-isopropyl N-dimethylphosphoramidate, triethyl phosphate and tetramethylphosphorodiamidic cyanide. The remaining impurities and byproducts constitute only trace amounts of the material present. Primary hydrolysis pathways of GA produce the major breakdown products of ethylphosphoryl cyanidate, dimethylamine, ethyl N, N-dimethylamidophosphoric acid, hydrogen cyanide, dimethylphosphoramidate and phosphoric acid. Hydrogen cyanide is a major hydrolysis product, but it is not unique to GA degradation.

Of these major degradation products, acute toxicity data are available for only a small number of the compounds. With respect to the major hydrolysis products, ethylphosphoryl cyanidate, dimethylamine and ethyl N, N-dimethylamidophosphoric acid may retain some of the parent compound's toxic properties. No toxicity studies have been publicly reported on these specific GA degradation products. Acute exposure data for the degradation products of GA have been described in the literature by Munro *et al.* and are given in Table 2.2 [5]. Although information is given for dimethylamine, cyanide and triethyl phosphate, it should be noted that these specific compounds are not unique to GA. Nonetheless, median lethal dose (LD_{50}) or median lethal concentration (LC_{50}) values are given for dimethylamine, diethyl dimethylphosphoramidate and triethyl phosphate. Lowest lethal dose (LD_{LO}) or lowest lethal concentration (LC_{LO}) values are only given for triethyl phosphate. Dimethylamine can be absorbed orally or by inhalation and is considered to be acutely toxic according to studies. In humans, severe burns can occur in the eye and on the skin with strong irritations to the olfactory system, similar to that of ammonia. Triethyl phosphate is also considered to be acutely toxic, with only moderate irritating properties. Chronic exposures to dimethylamine and triethyl phosphate have also been tabulated by Munro *et al.* in Table 2.3 [5]. Dimethylamine does not appear to be mutagenic or carcinogenic; however, triethyl phosphate has been well characterized [20]. Triethyl phosphate does seem to cause mutagenesis and affect the reproductive system [21].

2.2.2 Toxicity of GB (Sarin) Degradation Products

GB is the most volatile of the G-agents and is miscible with water, leaving water sources susceptible to contamination if large amounts are used. GB is also unstable and susceptible to hydrolysis. Stabilizers, such as dibutylchloramine,

Table 2.2 Acute exposure data for GA degradation products and impurities. Table taken from Reference [5] with permission from *Environmental Health Perspectives*

Compound	LD_{50} or LC_{50}	LD_{LO} or LC_{LO}	Additional effects
Dimethylamine	Rat: oral, 698 mg kg^{-1} [22] Mouse: oral, 316 mg kg^{-1} [22] Rabbit: guinea pig, oral, 240 mg kg^{-1} [22] Rat: inhalation, 4540 ppm per 6 h^{-1} [23] Mouse: inhalation, 7650 ppm per 2 h; 4725 ppm per 2 h [23, 24]	–	Human: nose, throat, lung irritation, 100 ppm [25] Human: eye, skin, severe burns Rabbit: eye, 50 mg per 5 min, severe irritation, opacity [26] Rat: sensory irritation, RD_{50}, 573 ppm [23] Mouse: sensory irritation, RD_{50}, 511 ppm [23]
Diethyl dimethylphosphor amidate	Mouse: IM, 440 mg kg^{-1} [27]	–	–
Triethyl phosphate	Rat: oral, 1311 mg kg^{-1} [28] Mouse: oral, 1500 mg kg^{-1} [29] Guinea pig: oral, 1600 mg kg^{-1} [30] Guinea pig: skin, >21 g kg^{-1} [28] Rat: inhalation, >2050 mg m^{-3} per 6 h, 28% respirable aerosol [28]	Rat: inhalation, 28 000 ppm per 6 h [31]	Rabbit: moderate eye irritation, 100 mg [28]

Table 2.3 Chronic exposure data for GA degradation products and impurities. Table taken from Reference [5] with permission from *Environmental Health Perspectives*

Compound	Carcinogenicity	Genetic effects	Reproductive effects	Systemic effects
Dimethylamine	Rat, mouse: inhalation negative (tentative) [25]	Mutagenicity: *Salmonella*, negative [32] Mutagenicity: CHO, negative [33] Rat: inhalation, cytogenetic, positive, aneuploidy and chromosome aberrations [34] CHO: chromosome aberrations, negative [33] CHO: sister chromatid exchange, negative [33] Cytogenicity: negative [35]	Rat: male, negative, 12 wk [36]	Rats, mice: decreased body weight gain; hematologic changes, inflammation and degeneration of olfactory epithelium [25]
Triethyl phosphate	–	Mutagenesis: positive, *Drosophila*, multiple doses [37–39] Mutagenesis: weakly positive, *Pseudomonas aeruginosa* [37] Mutagenesis: mostly negative, *Salmonella*, *Escherichia coli* [32, 40] Mutagenesis: negative, mouse, dominant lethal test [41] *In Vitro* transformation: negative, mouse cells [42]	Rat: oral, live birth index decrease, 57 g kg^{-1} (1% TEP) [21, 43]	Rat: oral, 120–150 d, increased liver weight, <10% TEP [21, 43] Rat: oral, 120–150 d, increased adrenal weight, 1, 5 and 10% TEP [21, 43] Rat: oral, 120–150 d, growth retardation, 5 and 10% TEP [21, 43]

tributylamine (TBA) and diisopropyl carbodiimide (DIPC), are added to prolong the lifetime of weapons-grade GB. Unlike GA, there are only a handful of degradation products for GB in addition to the stabilizers, mainly fluoride, isopropyl methylphosphonic acid (IMPA), methylphosphonic acid (MPA), diisopropyl methylphosphonic acid (DIMP) and methylphosphonic difluoride. IMPA, MPA and DIMP are very stable and persist for several months, if not longer. Fortunately, IMPA, MPA and DIMP possess only low toxicity concerns in both acute and chronic studies. Acute studies on methylphosphonic difluoride suggest that it is an irritant to the eyes and olfactory system and exhibits symptoms similar to that of a cholinesterase inhibitor [44]. Other stabilizers mentioned are moderately toxic with irritation to the eyes and olfactory system. Chronic exposures to the main degradation products of GB are not considered to be carcinogenic or mutagenic as noted in Table 2.4 [5].

2.2.3 Toxicity of GD (Soman) Degradation Products

The primary hydrolysis product of GD is pinacolyl methylphosphonic acid (PMPA). Other hydrolysis products include the breakdown of PMPA to MPA and pinacolyl alcohol. Major impurities found in GD consist of dipinacolyl methylphosphonate, methyl pinacolyl methylphosphonate, methyl methylphosphonofluoridate and methyl methacrylate. Toxicological data for MPA was discussed earlier in Section 2.2 and can be found in Table 2.4. There are no current toxicity data available on the major degradation product PMPA, although it is structurally similar to IMPA, which exhibits low toxicity (Tables 2.4 and 2.5). Toxicity reports have found that methyl methacrylate is considered to possess moderately acute effects and is irritating to the eyes

Table 2.4 Acute exposure data for GB degradation products and impurities. Table taken from Reference [5] with permission from *Environmental Health Perspectives*

Compound	LD_{50} or LC_{50}	LD_{LO} or LC_{LO}	Additional effects
Isopropyl methylphosphonic acid (IMPA)	Rat: oral, male, 7650 mg kg^{-1} [45] Rat: oral, female, 6070 mg kg^{-1} [45] Mouse: oral, male, 5620 mg kg^{-1} [45] Mouse: oral, female, 6550 mg kg^{-1} [45]	—	Rabbit: negative, eye irritant, 100 mg [45] Rabbit: mild skin irritant, 2 g kg^{-1} per 24 h [45]
Methylphosphonic acid (MPA)	Rat: oral, 5000 mg kg^{-1} [46] Mouse: oral, >5000 mg kg^{-1} [46]	—	Human: skin and eye irritant [47]
Diisopropyl methylphosphonic acid (DIMP)	Rat: oral, 826 mg kg^{-1} [48] Mouse: oral, 1041 mg kg^{-1} [48] Mink: oral, 503 mg kg^{-1} [49] Duck: oral, 1490 mg kg^{-1} [49] Cattle: oral, ~750 mg kg^{-1} [50]	—	Rat: ataxia, decreased activity; prostration, 430 mg kg^{-1} LOAEL (lowest observed adverse effect level) [48] Mouse: decreased activity; prostration, 430 mg kg^{-1} LOAEL [48] Mink: salivation, lethargy; immobilization, 300 mg kg^{-1} LOAEL [49]

Methylphosphonic difluoride	Mouse, dog: inhalation, 2700 mg m^{-3} per 30 min [51] Rat: inhalation, 8100 mg m^{-3} per 30 min [51] Monkey: inhalation, 3000 mg m^{-3} per 30 min [51] Guinea pig: inhalation, <1600 µg L^{-1} per 1 h (mg m^{-3}) [51]	–	Rat, mouse, dog, monkey: eye irritation [51] Rat, dog, monkey: corneal opacity, haze [51] Mouse: muscle weakness, ataxia [51] Dog, monkey: miosis [51] Rat, mouse, guinea pig: respiratory distress [44, 51] Dog: pulmonary edema, congestion [51]
Diisopropyl carbodiimide (DIPC)	Mouse: intravenous, 36 mg kg^{-1} [52]	–	–
Tributylamine (TBA)	Rat: oral, 740 mg kg^{-1} [53] Rat: oral, 540 mg kg^{-1} [54] Mouse: oral, 114 mg kg^{-1} [55] Rabbit: oral, 615 mg kg^{-1} [55] Guinea pig: oral, 350 mg kg^{-1} [55] Rabbit: skin, 250 µL kg^{-1} (194 mg kg^{-1}) [54]	Rat: inhalation, 75 ppm per 4 h [54]	Human: central nervous system stimulation, skin irritation, sensitization [56, 57] Human: eye, skin, and respiratory irritant [58]

Table 2.5 Chronic exposure data for GB degradation products. Table taken from Reference [5] with permission from *Environmental Health Perspectives*

Compound	Carcinogenicity	Genetic effects	Reproductive effects	Systemic effects
Isopropyl methylphosphonic acid (IMPA)	–	Mutagenicity: *Salmonella*, negative [45]	–	No toxicity to rats fed 300 ppm in water for 90 d [45]
Diisopropyl methylphosphonic acid (DIMP)	–	Mutagenicity: *Salmonella*, negative [59] Mutagenicity: *S. cerevisiae*, negative [59]	Mink: oral, negative, two-generation study [60] Rat: oral, negative, three-generation study [59] Mink: oral, negative, one-generation study [49]	Mink: mild hematologic effects [60] Guinea pig: dermal hypersensitivity, negative [48]

and skin (Table 2.6). Studies containing data for chronic exposures to methyl methacrylate have indicated mutagenic and reproductive toxicity (Table 2.7).

2.2.4 Toxicity of GF (Cyclosarin) Degradation Products

GF is similar to GB in structure. The United States did not produce GF in significant amounts; therefore, its chemistry and toxicity have been less explored, and the case is similar for the degradation products. The primary hydrolysis products of GF include hydrofluoric, cyclohexylmethylphosphonic, methylphosphonic acids as well as cyclohexanol. These degradents are less toxic than the parent compound GB. Most GF degradation products are similar to the situation for GB. As GF is similar in structure, most of the degradation compounds are similar to that of Sarin, mainly MPA (Table 2.4), but cyclohexylmethylphosphonic acid and cyclohexanol toxicity studies have not been not reported. It is believed that cyclohexanol is acutely toxic according to the Pesticide Action Network (PAN) pesticides database [78].

2.2.5 Toxicity of VX Degradation Products

VX is less volatile than the G-agents and is resistant to hydrolysis. Degradation of VX can occur by several different pathways [18, 79–82]. There are over 30 degradation products of VX [18, 79–86], but one of the major and well known degradation products is S-(2-diisopropylaminoethyl) methylphosphonothioic acid (or EA 2192), which is highly soluble in water and possesses toxic properties similar to those of the parent compound. Other common degradation products of VX include ethyl

Table 2.6 Acute exposure data for GD degradation products and impurities. Table taken from Reference [5] with permission from *Environmental Health Perspectives*

Compound	LD_{50} or LC_{50}	LD_{LO} or LC_{LO}	Additional effects
Methylphosphonic acid (MPA)	Rat: oral, 5000 mg kg^{-1} [46] Mouse: oral, >5000 mg kg^{-1} [46]	–	Human: skin and eye irritant [47]
Methylphosphonic difluoride	Mouse, dog: inhalation, 2700 mg m^{-3} per 30 min [51] Rat: inhalation, 8100 mg m^{-3} per 30 min [51] Monkey: inhalation, 3000 mg m^{-3} per 30 min [51] Guinea pig: inhalation, <1600 μg L^{-1} per 1 h (mg m^{-3}) [44]	–	Rat, mouse, dog, monkey: eye irritation [51] Rat, dog, monkey: corneal opacity, haze [51] Mouse: muscle weakness, ataxia [51] Dog, monkey: miosis [51] Rat, mouse, guinea pig: respiratory distress [44, 51]
Methyl methacrylate	Rat: oral, 7872 mg kg^{-1} [61] Mouse: oral, 3625 mg kg^{-1} [62] Dog: oral, 4725 mg kg^{-1} [63] Rabbit: oral, 8700 mg kg^{-1} [62] Guinea pig: oral, 5954 mg kg^{-1} [63] Rat: inhalation, 78 000 mg m^{-3} per 4 h [64] Mouse: inhalation, 18 500 mg m^{-3} per 2 h [64] Rabbit: skin, >5 g kg^{-1} [65]	Dog: inhalation, 41 200 mg m^{-3} per 3 h [63] Rabbit: inhalation, 17 500 mg m^{-3} per 4 h [61] Guinea pig: inhalation, 19 000 mg m^{-3} per 4 h [61]	Rabbit: skin, irritation, 10 g kg^{-1} [61] Rabbit: eye, 150 mg [63]

Table 2.7 Chronic exposure data for GD impurity. Table taken from Reference [5] with permission from *Environmental Health Perspectives*

Compound	Carcinogenicity	Genetic effects	Reproductive effects	Systemic effects
Methyl methacrylate	Mouse: rat, inhalation, negative [66] Rat: oral, negative [67]	Mutagenicity: positive, *Salmonella* [68] Mutagenicity: Ames, negative [32] Mutagenicity: positive, mouse lymphoma cells [69] Micronucleus test: equivocal, mouse lymphocyte [69] Mutation: mouse lymphoma cells, positive, with activation [70] Cytogenetic effects: positive, several end points [66, 69, 70] SCE: positive; hamster ovary [66] Mouse: negative, RDS test [71] hamster ovary Mouse: negative, RDS test [71]	Rat: inhalation, fetotoxicity, growth retardation; embryolethality [72] Rat: inhalation, postimplantation mortality [73]	Human: contact dermatitis [74] Rat: inhalation, serum composition, cholesterol, bilirubin, transaminases, nutrition/metabolic effects [75] Rat: inhalation, focal lung fibrosis, olfactory sensory epithelium degeneration, nasal cavity inflammation [76] Rat: oral, somnolence, neurostructural changes, lipid effects [77] Mouse: inhalation, zonal hepatocellular necrosis, olfactory sensory epithelium degeneration, nasal cavity inflammation, epithelial hyperplasia [66]

methylphosphonic acid (EMPA), MPA, diisopropylamin-oethane thiol (DESH), bis(diisopropylaminoethyl) disul-fide (EA 4196), bis(diisopropylaminosulfide), numerous diisopropylaminoethane compounds, including 1,9-bis (diisopropyl-amino)-3,4,7-trithianonane, N, N'-methane-tetrayl bis(cyclohexanamine) and mono(2-ethylhexyl)ester hexanedioic acid. Acute toxicity data of the degradation products of VX are listed in Table 2.8 with several compounds, such as EA 2192, believed to retain their anticholinesterase activity and toxicity. Most of the degradation products do not have sufficient or adequate toxicity data; however, chronic exposure data for some of the degradation products and stabilizers are listed in Table 2.9.

2.3 TOXICITY OF BLISTER AGENTS

Lewisite, when used as a chemical warfare agent, affects the eyes, respiratory tract and skin, these being the parts most exposed to the gas. Furthermore, Lewisite has lipophilic properties and penetrates easily through the skin [120].

According to data presented by Colonel Richard Solana of the US Army Medical Research Institute of Chemical Defense, the LD_{50} for humans is estimated to be 40 mg kg^{-1} for liquid Lewisite, and 100 000 mg-min m^{-3} for Lewisite vapor. The LD_{50} often quoted for liquid Lewisite is con-sidered low by many investigators [78]. Lewisite lethality in humans, when delivered via the skin, is also related to the physical state on delivery. In the case of dermal expo-sure, $0.05-0.1$ mg cm^{-2} produces erythema, 0.2 mg cm^{-2} produces vesication and $\sim$30 drops (2.6 mg) can kill an average man through systemic toxicity. In the case of ocu-lar exposure, 15 min exposure to a vapor concentration of 10 mg m^{-3} produces conjunctivitis. It is assumed that upon

Table 2.8 Acute exposure data for VX degradation products and impurities. Table taken from Reference [5] with permission from *Environmental Health Perspectives*

Compound	LD_{50} or LC_{50}	LD_{LO} or LC_{LO}	Additional effects
Diisopropyl ethyl mercaptoamine (DESH)	Mouse: ip, 5 mg kg^{-1} [87]	–	–
S-(Diisopropylaminoethyl) methylphosphonothionate (EA 2192)	Rat: oral, 630 µg kg^{-1} [88] Rat: iv, 18 µg kg^{-1} [88] Mouse: iv, 50 µg kg^{-1} [89] Rabbit: iv, 0.017 mg kg^{-1} [90] Rabbit: iv, 12 µg kg^{-1} [89]	–	–
2-Diisopropylaminoethanol	Rat: oral, 860 mg kg^{-1} [91] Rat: oral, 1070 mg kg^{-1} [92] Mouse: oral, 770 mg kg^{-1} [91] Rabbit: skin, 450 µL kg^{-1} [92] Rat: inhalation, 1965 mg m^{-3} per 6 h [91] Mouse: inhalation, 1661 mg m^{-3} per 6 h [91]	–	Rabbit: skin irritant, severe corrosive [91] Rabbit: skin irritant, mild [92] Rabbit: skin irritant, 500 mg open, mild [93] Rabbit: eye irritant, 750 µg open, severe [92]
Diethyl methylphosphonate	Mouse: ip, 2240 mg kg^{-1} [94]	–	–
Methylphosphonic acid (MPA)	Rat: oral, 5000 mg kg^{-1} [46] Mouse: oral, >5000 mg kg^{-1} [46]	–	Human: skin and eye irritant [47]

Table 2.8. (*continued*)

Compound	LD$_{50}$ or LC$_{50}$	LD$_{LO}$ or LC$_{LO}$	Additional effects
Diethyl dimethyl pyrophospho-nate	Rabbit: skin, 7.1 mg kg^{-1} [90]	–	–
O,S-diethyl methylphos-phonothioate	Rat: oral, 6.0 mg kg^{-1} [95] Dog: iv, 5620 µg kg^{-1} [96] Mouse: iv, 1 mg kg^{-1} [96] Rabbit: iv, 2480 µg kg^{-1} [96]	–	–
Diisopropyl-amine	Rat: oral, 770 mg kg^{-1} [92] Mouse: oral, 2120 mg kg^{-1} [97] Rabbit: oral, 4700 mg kg^{-1} [97] Guinea pig: oral, 2800 mg kg^{-1} [97] Rabbit: skin, >10 g kg^{-1} [98]	Rat: inhalation, 4800 mg m^{-3} per 2 h [99] Rat: inhalation, <2200 ppm per 3 h; >777 ppm per 7 h [100] Mouse: inhalation, 4200 mg m^{-3} per 2 h [99] Cat: inhalation, 2207 ppm per 72 min [101] Rabbit: inhalation, 2207 ppm per 2.5 h [101] Guinea pig: inhalation, 2207 ppm per 82 min [101]	Human: vision disturbances, nausea and headache, 25–50 ppm [101, 102] Several species: cloudy swelling of corneal epithelium, >600 ppm [101] Rabbit: mild skin irritant, 500 mg per 24 h [103] Guinea pig: severe skin irritant, 3 wk [104] Rabbit: severe eye irritant, 750 µg open [92]

Diisopropyl-carbodiimide	Mouse: iv, 36 mg kg^{-1} [52]	—	Human: severe eye irritation, temporary blindness [105]
Dicyclohexyl-carbodiimide	Rat: oral, ~400 mg kg^{-1} [106] Mouse: oral, >800 mg kg^{-1} [106] Guinea pig: dermal, 1–5 drops [106] Rat: inhalation, 0.159–0.417 mg L^{-1} [106]	—	Guinea pig: skin, moderate irritant [106] Rabbit: eye, severe irritant [106] Rat: inhalation, lung inflammation, focal atrophy of stomach, liver necrosis, testicular atrophy [106]

Table 2.9   Chronic exposure data for VX degradation products and stabilizers. Table taken from Reference [5] with permission from *Environmental Health Perspectives*

Compound	Carcinogenicity	Genetic effects	Systemic effects
Diisopropyl-amine	–	Mutagenicity: negative, Ames [107] Mutagenicity: questionably positive, Ames, 1 μg per plate [102, 108] DNA repair: negative, rat hepatocyte primary culture assay [104]	Guinea pig: sensitizer, negative [104]
Diisopropyl-carbodiimide	Carcinogenicity: negative, 20-wk prechronic studies, female mice [109, 110]	Mutagenicity: negative, Ames [111] Cytogenicity: negative, micronucleus induction, Fischer 344 rat [112] Cytogenicity: positive, micronucleus induction, B6C3F1 mouse [112]	Human: contact allergen [113] Mouse: skin sensitizer [114]
Dicyclohexyl-carbodiimide	Mouse: likely animal Carcinogen [115, 116]	Mutagenicity: negative, Ames [117] Cytogenicity: positive, micronucleus, B6C3F1 mice [118] Cytogenicity: negative, micronucleus, Fischer 344 rat [118]	Human: contact allergen [114] Mouse: skin sensitizer [119]

entry of Lewisite into the aqueous medium of the intact skin it is rapidly hydrolyzed to a stable, water-soluble, but highly toxic derivative, 2-chlorovinylarsine oxide (Lewisite oxide), and hydrochloric acid. Feister and colleagues (1989) postulated that Lewisite oxide may be the principal metabolite and major cytotoxic form of Lewisite within tissues. It is also believed that the trivalent form of arsenic, which is highly reactive in biological systems, is responsible for the overt toxicity of all arsenicals, including Lewisite, to living systems [121, 122].

Although Lewisite is only slightly soluble in water [0.5 g L^{-1}] [123], hydrolysis is rapid and results in the formation of the water-soluble dihydroxy arsine (2-chlorovinyl arsonous acid). In basic solution, the *trans*-Lewisite isomer is cleaved by the hydroxyl ion to give acetylene and sodium arsenite; this occurs even at low temperatures [79, 123]. *cis*-Lewisite must be heated to over 40°C to react with NaOH to yield vinyl chloride, sodium arsenite and acetylene. In aqueous solution, the *cis*-isomer undergoes a photoconversion into the *trans*-isomer. In water and in the presence of oxidizers naturally present in the environment, the toxic trivalent arsenic of Lewisite oxide is oxidized to the less toxic pentavalent arsenic [124]. Regardless of the degradation pathway, arsenical compounds will ultimately be formed.

The effects of acute exposure to Lewisite degradation products are given in Table 2.10 [5]. When Lewisite 1 reacts with water, an arsenic based compound will form: chlorovinyl arsonous acid (CVAA) and hydrochloric acid (HCl), see Equation 2.1:

$$ClCH=CHAsCl_2 + 2H_2O \rightleftarrows ClCH=CHAs(OH)_2$$
$$+ 2HCl \quad (2.1)$$

Table 2.10 Effects of acute exposure to Lewisite degradation products [5]. Table taken from Reference [5] with permission from *Environmental Health Perspectives*

Degradation product Lewisite	LD$_{50}$ median lethal dose	LD$_{LO}$ lowest lethal dose
Chlorovinyl arsenous oxide C$_2$H$_4$AsClO	Mouse: subcutaneous, 5 mg kg^{-1} [125]	–
2-Chlorovinyl arsonic acid C$_2$H$_4$AsClO$_3$	–	Rat: oral, 50 mg kg^{-1} [126]

In a situation with water abundance, the reaction proceeds to form chlorovinyl arsonous oxide (CVAO), see Equation 2.2:

$$ClCH{=}CHAs(OH)_2 \rightleftharpoons ClCH{=}CHAsO + 2H_2O \qquad (2.2)$$

CVAO, also known as Lewisite oxide, is more stable than CVAA, therefore this reaction is preferred (Equation 2.2) [127].

Nitrogen Mustard agents produce blistering on the skin and wounds resembling burns. Blistering not only happens on the skin but it can affect other organs as well. Blistering in the lungs can produce bleeding and can cause respiratory problems. Other body parts that can be affected are: scrotum, face, anus, legs, buttocks, hands and feet. The leading cause of death after Nitrogen Mustard exposure is lung injury, which starts with mild symptoms and can end up with pulmonary edema [128].

Sulfur Mustard is an alkylating agent. Alkylating agents bind covalently to various nucleophilic molecules such as DNA, RNA, proteins and components of cell membranes.

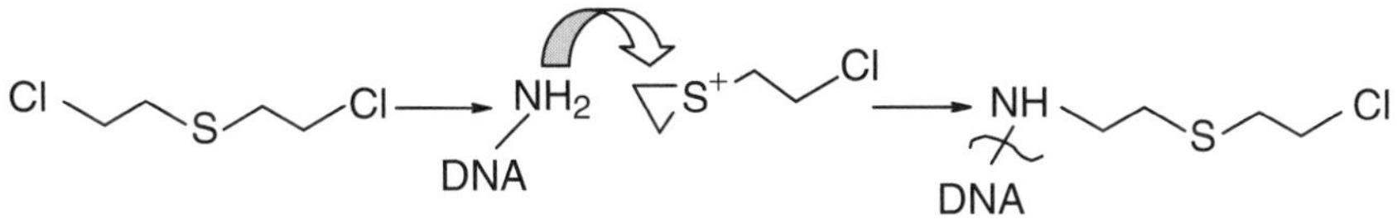

Figure 2.2   Cross-linking of DNA strands caused by Sulfur Mustard

Sulfur Mustard can cause cross-linking of DNA strands (Figure 2.2), alkylation of RNA molecules, altered translation and altered protein synthesis leading to cell death. Binding to proteins mainly with the thiol group of cysteine produces structural changes, altering normal cell physicology, specifically enzyme activity. Relevant animal studies have shown the following results: LD_{50}, through skin contact, rat 9, dog 20 and rabbit 100 mg kg^{-1}; LC_{50}, by inhalation (for 10 min), rat 100, rabbit 280 and monkey 80 mg m^{-3} [129].

Thiodiglycol is the main hydrolysis product of Sulfur Mustard and is a Schedule 2 compound. Limited information is known about Thiodiglycol, and few toxicological studies have been performed on this degradation product. The oral LD_{50} values were >5000 mg kg^{-1} in rats [130]. Other Sulfur Mustard degredants include BHET-alkenes (Figure 2.3) with little known about their toxicity.

The LC_{50} is approximately 1500 mg-min m^{-3} for HN-1 and HN-3, and 3000 mg-min m^{-3} for HN-2. Exposure to Nitrogen Mustard, even at low concentrations, can also affect eyes, skin and mucous membranes. The International Agency for Research on Cancer (IARC) has classified Nitrogen Mustard as probably being carcinogenic to humans (Group 2A). There is some evidence that it causes leukemia in humans, and it has been shown to cause leukemia and cancers of the lung, liver, uterus and large intestine in animals [131].

On the basis of acute studies with laboratory animals, methyldiethanolamine is considered slightly toxic by a

Compound name (acronym)	Structure

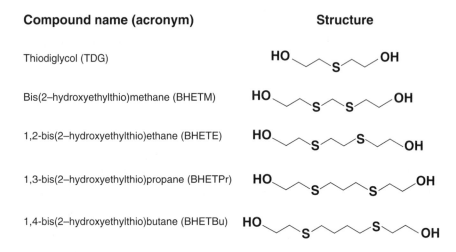

Thiodiglycol (TDG)

Bis(2–hydroxyethylthio)methane (BHETM)

1,2-bis(2–hydroxyethylthio)ethane (BHETE)

1,3-bis(2–hydroxyethylthio)propane (BHETPr)

1,4-bis(2–hydroxyethylthio)butane (BHETBu)

Figure 2.3   Hydrolysis degradation products of Sulfur Mustard

single oral dose and practically nontoxic by a single dermal application. The oral LD_{50} value in the rat is $4.78\,g\,kg^{-1}$ and the dermal LD_{50} value in the albino rabbit is $6.24\,g\,kg^{-1}$.

Nitrogen Mustards undergo hydrolysis leading to three alkyl ethanolamines: N-methyldiethanolamine (MDEA), N-ethyldiethanolamine (EDEA) and triethanolamine (TEA), [degradation/precursor products of: bis(2-chloroethyl)ethyl amine – Nitrogen Mustard-1, HN-1; methyl-bis-(2-chlor oethyl)amine – Nitrogen Mustard-2, HN-2; and tris(2-chloroethyl)amine – Nitrogen Mustard-3, HN-3 blister agents] [132].

Methyldiethanolamine is considered moderately irritating to the eyes, but only slightly irritating to the skin. The product is not corrosive under the conditions of the US Department of Transportation (DOT) corrosivity test and is not regulated as a hazardous material for transportation purposes [133].

Phosgene Oxime is toxic through inhalation, ingestion or skin contact. The effects of the poisoning occur almost immediately. No antidote for Phosgene Oxime poisoning is known.

Inhaled Phosgene Oxime is extremely irritating to the upper airways and causes pulmonary edema. Irritation occurs with exposures to 0.2 mg-min m^{-3} and becomes unbearable at 3 mg-min m^{-3}. The estimated LCt$_{50}$ (the product of concentration times the time that it is lethal to 50% of the exposed population by inhalation) is $1500–2000$ mg-min m^{-3}. Pain and local tissue destruction occur immediately on contact with skin, eyes and mucous membranes. Phosgene Oxime is rapidly absorbed from the skin and eyes and may result in systemic toxicity. The LD$_{50}$ for skin exposure is estimated to be 25 mg kg^{-1}. Regarding ingestion, no human data is available. Animal studies suggest Phosgene Oxime may induce hemorrhagic inflammatory lesions in the gastrointestinal tract [131].

2.4 TOXICITY OF STERNUTATOR AGENTS

Diphenylcyanoarsine (Clark I) (DA), Diphenylchloroarsine (Clark II) (DC) and Adamsite (DM) are all classified as sternutator agents because they cause extreme sneezing, coughing and vomiting. All three chemicals have very similar structures (Figure 2.4) and all contain arsenic; however, the levels that cause such a violent reaction are not toxic (0.1 mg m^{-3}). Once the particles fall to the ground after dispersion, they become ineffective unless they are re-dispersed. To have any effect, the aerosols must be inhaled or absorbed by the mucous of the eyes.

Exposure to these agents is hard to detect due to the lack of odor or color of the smoke at low concentrations. Once exposed, the onset of symptoms is slow and can last from 30 min to a couple hours. Symptoms include uncontrollable sneezing, coughing, headache and vomiting [134]. Although

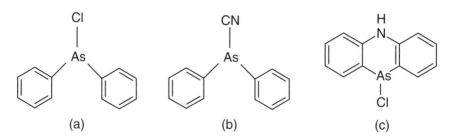

Figure 2.4 Structure of: (a) Clark I (DC); (b) Clark II (DA); and (c) Adamsite (DM)

not well known, the agents are believed to work by inhibiting SH-containing enzymes. This inhibition interferes with the functioning of a cell's metabolic processes and results in destruction of the cell structure [135]. The treatment of symptoms includes using antiemtic therapies.

Currently, toxicity studies for the parent agents are rare. The group working with Henriksson [136] has studied the toxicity *in vitro* and *in vivo* using flow cytometry. The effects of cell proliferation and cell kinetics were explored in both the *in vitro* and *in vivo* situations. The researchers concluded that the chemical warfare agents were more potent with regards to cell proliferation than arsenic pentoxide (As_2O_5). This shows that the arsenic ion only plays a partial role in the toxicity of these agents.

2.4.1 Toxicity of Degradation Products of Sternutator Agents

The formation of the degradation products usually occurs from slow hydrolysis in water. At the end of World War II, tons of chemical warfare agents were dumped into the sea by both Germany and Japan. Although Clark I and II are not easily soluble in water, over time they will begin to break down into toxic products containing arsenic. There are eight degradation products from Clark I and Clark II and their degradation pathways are shown in Figure 2.5. The main products are diphenylarsinic acid (DPAA), phenylarsonic acid (PAA), phenylarsinic oxide (PAO), triphenylarsine (TPA) and triphenylarsine oxide (TPAO). The degradation products are considered to be more toxic than the parent agents.

The main focus is on diphenylarsinic acid (DPAA). DPAA and Dimethylarsinic acid (DMA) [As(V)] (dimethylarsinic acid) have been shown to be toxic to human cells when interacting with SH compounds. Researchers saw an increase in cytotoxicity as well as structural changes within the cells [137]. Kroening and coworkers also studied the effects of cell death by the degradation products and were able to conclude that PAO, PAA, DPAA, TPA and TPAO are more toxic than common arsenic standards [138].

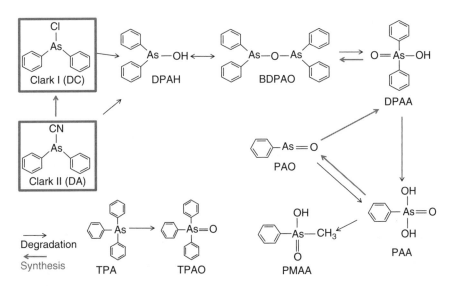

Figure 2.5 Degradation pathway for Clark I and Clark I. Figure courtesy of http://www.dstl.gov.uk/conferences/cwd/2007/pres/ hanaoka-pres.pdf

Other studies have focused on the effects of the dumped chemical warfare agents on marine life and also on the people who use the water. It was noted in Kamisu, Japan, that certain residents had the onset of symptoms including staggering, hand tremors and even mental disabilities in young children. Researchers were able to trace the cause to DPAA present in the well water [139]. Mice were given DPAA orally in attempts to 'recreate' what happened with the residents and the well water. Cerebral symptoms were also seen in the mice allowing researchers to understand the oxidative stress that was taking place in the brain due to DPAA poisoning [140]. Risk assessments have also been done on the fish present at dumping sites in the Baltic Sea [141, 142]. The highest risk degradation product for arsenic-containing chemical warfare agents is TPA, followed by Adamsite and Clark I. These chemicals when ingested orally, pose high carcinogenic and mutagenic risks. Even with risk assessment, more studies are needed to investigate the risks to fish and from the consumption of fish.

REFERENCES

[1] Epstein, J., Rosenblatt, D. H., Gallacio, A., McTeague, W. F. (1973), Summary report on a data base for predicting consequences of chemical disposal operations, EASP 1200-12, January 1973, AD-B955399 (distribution limited to U.S. Government), *EASP 1200-12 1973*, AD-B955399 (distribution limited to U.S. Government).

[2] Chauhan, S. Chauhan, S., D'Cruz, R. *et al.* (2008) *Environ. Toxicol. Pharmacol.*, **26**, 113–122.

[3] Burdon, J. (2007) *Molecules of Death*, 2nd edn, (eds R.H. Waring, G.B. Steventon and C. Mitchell), Imperial College Press, London, pp. 209–231.

[4] Technical Secretariat of the Organisation for Prohibition of Chemical Weapons (1997) *Convention on the Prohibition of the*

Development, Production, Stockpiling and Use of Chemical Weapons and on their Destruction, The Hague, http://www.opcw.nl (accessed 10 June 2010).

[5] Munro, N.B., Talmage, S.S, Griffin, G.D. *et al.* (1999) *Environ. Health Persp.*, **107**, 933–974.

[6] Bajgar, J. Fusek, J., Kassa, J., *et al.* (2009) *Curr. Med. Chem.*, **16**, 2977–2986.

[7] Jokanovic, M. (2009) *Toxicol. Lett.*, **188**, 1–10.

[8] Raber, E. Jin, A., Noonan, K. *et al.* (2001) *Int. J. Environ. Health Res.*, **11**, 128–148.

[9] Dabisch, P.A., Hulet, S.W., Kristovich, R., Mioduszewski, R.J. (2008), in *Chemical Warfare Agents Chemistry, Pharmacology, Toxicology, and Therapeutics*, 2nd edn (eds J.A. Romano, B.J. Lukey, and H. Salem), CRC Press, Baco Raton, pp. 233–246.

[10] Center for Disease Control and Prevention (2010) *Emergency Preparedness and Response: Toxic Syndrome Description*, http://www.bt.cdc.gov/agent/nerve/tsd.asp (accessed 17 June 2010).

[11] Respirex (2010) *Personal and Respiratory Protective Equipment*, http://www.respirex.co.uk (accessed 17 June 2010).

[12] Bajgar, J., Kuca, K., Jun, K. *et al.* (2007) *Curr. Drug Metab.*, **8**, 803–809.

[13] van der Schans, M.J., Benschop, H.P., Whalley, C.E. (2008), in *Chemical Warfare Agents Chemistry, Pharmacology, Toxicology, and Therapeutics*, 2nd edn (eds J.A. Romano, B.J. Lukey, and H. Salem), CRC Press, Baco Raton, pp. 97–122.

[14] Creasy, W.R., Stuff, J.R., Williams, B. (1997) *J. Chromatogr. A*, **774**, 253–263.

[15] D'Agostino, P.A., Hansen, A.S., Lockwood, P.A., Provost, L.R. (1985) *J. Chromatogr.*, **347**, 257–266.

[16] D'Agostino, P.A., Provost L.R. (1992) *J. Chromatogr.*, **589**, 287–294.

[17] D'Agostino, P.A. Provost, L.R. (1992) *J. Chromatogr.*, **589**, 89–95.

[18] D'Agostino, P.A., Provost, L.R., Looye, K.M. (1989) *J. Chromatogr.*, **465**, 271–283.

[19] MacNaughton, M.G., Brewer, J.H. (1994) *Environmental Chemistry and Fate of Chemical Warfare Agents*, Southwest Research Institute, San Antonio, TX, Document number: Project 01-5864.

[20] U.S. EPA (1985) *Draft Report: Chemical Hazard Information Profile, Triethyl Phosphate*, U.S. Environmental Protection Agency, Washington, DC, Document 40-8542937.
[21] Gumbmann, R.R., Gagne, W.E., Williams, S.N. (1968) *Appl. Pharmacol.*, **12**, 360–371.
[22] Dzhanashvili, G.D. (1967) *Hyg. Sanit.*, **32**, 329–355.
[23] Steinhagen, W.H., Swenberg, J.A., Barrow, C.S. (1982) *Am. Ind. Hyg. Assoc. J.*, **43**, 411–417.
[24] Mezentseva, N.V. (1956) *Gig. Sanit.*, **21**, 47–49.
[25] Rothwell, C.E., Turck, P., Parker, D., Rowland, J., England, T. (1990) *Summary Review of Health Effects Associated with Dimethylamine: Health Issue Assessment*, U.S. Environmental Protection Agency, Research Triangle Park, NC.
[26] Mellerio, J., Weale, R.A. (1966) *Br. J. Ind. Med.*, **23**, 153–154.
[27] Grechkin, N.P., Grishina, L.N., Neklesova, I.D. *et al.* (1977) *Pharm. Chem. J.*, **11**, 38–41.
[28] Eastman Kodak Company (1984) *Material Safety Data Sheet*, Office of Toxic Substances, U.S. Environmental Protection Agency, Washington, DC, MSDS-10, 100A-01 (010-84), FYI-OTS-0884-0328 Supp, Seq. E.
[29] Bisesi, M.S. (1994) in *Patty's Industrial Hygiene and Toxicology*, 4th edn (eds G.D. Clayton and F.E. Clayton), John Wiley & Sons, Inc., New York, pp. 2967–3118.
[30] Deichman, W.B., and Gerarde, H.W. (1969) *Toxicology of Drugs and Chemicals*, Academic Press, New York.
[31] Levin, L., Gabriel, K.L. (1973) *Am. Ind. Hyg. Assoc. J.*, **34**, 286–291.
[32] Zeiger, E., Anderson, B., Haworth, S. *et al.* (1987) *Environ. Mutagen.*, **9**, 1–110.
[33] Hsie, A.W., San Sebastian, J.R., Perdue, S.W. *et al.* (1987) *Mol. Toxicol.*, **1**, 217–234.
[34] Isakova, G.K., Yakshtat, B.Y., Kerkis, Y.Y. (1971) *Gig. Sanit.*, **36**, 9–13.
[35] Ishidate, M.J., Sofuni, T., Yoshikawa, K. (1981) *Gann. Monogr. Cancer Res.*, **27**, 95–108.
[36] Jackson, H., Craig, A.W. (1966) *Nature*, **212**, 86–87.
[37] Dyer, K.F., Hanna, P.J. (1973) *Mutat. Res.*, **21**, 175–177.
[38] Graf, U., Frei, H., Kagi, A. *et al.* (1989) *Mutat. Res.*, **222**, 359–373.

[39] Hanna, P.J., Dyer, K.F. (1975) *Mutat Res.*, 28, 405–420.
[40] U.S. EPA Chemical Hazard Information Profile (1985) *Draft Report: Triethyl Phosphate*, U.S. Environmental Protection Agency, Washington, DC, EPA/OTS Doc 40-8542937.
[41] Epstein, S.S., Arnold, E., Andrea, J. *et al.* (1972) *Toxicol. Appl. Pharmacol.*, 23, 288–325.
[42] Eastman Kodak Company (1984) *Material Safety Data Sheet: Basic Toxicity, Bacterial Mutagenicity, and in vitro Mammalian Cell Transformation of Triethyl Phosphate*, Office of Toxic Substances, U.S. Environmental Protection Agency, Washington, DC, MSDS-10, 100A-01 (010-84), FYI-OTS-0884-0328 Supp, Seq. E.
[43] Gumbmann, R.R., Gagne, W.E., Williams, S.N. (1968) *Toxicol. Appl. Pharmacol.*, 12, 360–371.
[44] Dahl, A.R., Marshall, T.C., Hobbs, C.H. (1984) *Toxicologist*, 4, 22.
[45] Mecler, F.J. (1981) *Mammalian Toxicological Evaluation of DIMP and DCPD (Phase 3-IMPA)*, U.S. Army Medical Research and Development Command, Fort Detrick, MD.
[46] Williams, R.T., Miller, W.R.I., MacGillivray, A.R. (1987) *Environmental Fate and Effects of Tributyl Phosphate and Methyl Phosphonic Acid*, U.S. Army Armament Munitions Chemical Command, Chemical Research, Development and Engineering Center, Aberdeen Proving Ground, MD.
[47] Thiokol/Ventron, (1980) *Material Safety Sheet: Methylphosphonic Acid*, Thiokol/Ventron Division, Danvers, MA.
[48] Hart, E.R. (1976) *Mammalian Toxicological Evaluation of DIMP and DCPD. Final Report*, Litton Bionetics, Inc., Kensington, MD.
[49] Aulerich, R.J., Coleman, T.H., Polin, D. *et al.* (1979) *Toxicology Study of Diisopropyl Methylphosphonate and Dicyclopentadiene in Mallard Ducks, Bobwhite Quail and Mink*, Michigan State University, East Lansing, MI.
[50] Palmer, J.S., Cysewski, S.J., Crockshank, H.R. *et al.* (1979) *Toxicologic Evaluation and Fate of Diisopropyl Methylphosphonate (DIMP) and Dicyclopentadiene (DCPD) in Cattle*, National Technical Information Service, Springfield, VA.
[51] Crook, J.W., Musselman, N.P., Hess, T.L., Oberst, F.W. (1969) *Toxicol. Appl. Pharmacol.*, 15, 131–135.

[52] Registry of Toxic Effects of Chemical Substances Database (1997) *Diisopropylcarbodiimide*, MEDLARS Online Information Retrieval System, National Library of Medicine. Bethesda, MD.
[53] Ciugudeanu, M., Gabor, S., Gocan, M. *et al.* (1985) *Rev. Chim.*, **36**, 667–670.
[54] Carpenter, C.P., Weil, C.S., Smyth, H.F.J. (1974) *Toxicol. Appl. Pharmacol.*, **28**, 313–319.
[55] Din Min, L. (1977) *Gig. Sanit.*, **42**, 36–38.
[56] Budavari, S., O'Neil, M.J., Smith, A. (1996), in *The Merck Index. An Encyclopedia of Chemicals, Drugs, and Biologicals*, 12th edn (ed. P.E. Heckelman), Merck & Co., Inc., Rathway, NJ.
[57] Lewis, R.J.S. (1996) *Sax's Dangerous Properties of Industrial Materials*, Van Nostrand Reinhold, New York.
[58] Plaa, G. (1978) *Toxicology as a Predictive Science*, Proceedings of the 1st International Congress on Toxicology, New York (ed. W. Duncan), Academic Press.
[59] Hart, E.R. (1980) *Mammalian Toxicological Evaluation of DIMP and DCPD (Phase II)*, Litton Bionetics, Inc., Kensington, MD.
[60] Bucci, T.J., Mercieca, M.D., Perman, V., Weiss, D.J. (1997) *Two-Generation Reproductive Study in Mink Fed Diisopropyl Methylphosphonate (DIMP). Final Report*, Pathology Associates International, Frederick, MD.
[61] Deichman, W.B. (1941) *J. Ind. Hyg. Toxicol.*, **23**, 343–351.
[62] Klimkina, N.V., Ekhina, R.S., Sergeev, A.N. (1976) *Gig. Sanit.*, **41**, 6–11.
[63] Spealman, C.R., Main, R.J., Haag, H.B., Larson, P.S. (1945) *Ind. Med.*, **14**, 292–298.
[64] Blagodatin, V.M., Smirnova, E.S., Dorofeeva, E.D. *et al.* (1976) *Gig. Tr. Prof. Zabol.*, **6**, 5–8.
[65] Registry of Toxic Effects of Chemical Substances Database (1997) *Methyl methacrylate*, MEDLARS Online Information Retrieval System, National Library of Medicine, Bethesda, MD, Document number: NTIS OTS0520933.
[66] National Toxicology Program, (1986) *Toxicology and Carcinogenesis Studies of Methyl Methacrylate (CAS No. 80-62-6) in F344/N Rats and B6C3F1 Mice (Inhalation Studies)*, National Toxicology Program, Research Triangle Park, NC.
[67] Borzelleca, J.F., Larson, P.S., Hennigar, G.R.J. *et al.* (1964) *Toxicol. Appl. Pharmacol.*, **6**, 29–36.

[68] Ashby, J., Tennant, R.W. (1991) *Mutat. Res.*, **257**, 229–306.
[69] Doerr, C.L., Harrington-Brock, K., Moore, M.M. (1989) *Mutat. Res.*, **222**, 191–203.
[70] Amtower, A.L., Brock, K.H., Doerr, C.L. *et al.* (1986) *Environ. Mutagen.*, **8**, 4.
[71] Miyagawa, M., Takasawa, H., Sugiyama, A. *et al.* (1995) *Mutat. Res.*, **343**, 157–183.
[72] Nicholas, C.A., Lawrence, W.H., Autian, J. (1979) *Toxicol. Appl. Pharmacol.*, **50**, 451–458.
[73] Luo, S.-Q., Gang, B.-Q., Sun, S.-Z. (1986) *Toxicol. Lett.*, **31**, 80.
[74] Kassis, V., Vedel, P., Darre, E. (1984) *Contact Dermatitis*, **11**, 26–28.
[75] Tansy, M.F., Kendall, F.M., Benhayem, S. *et al.* (1976) *Environ. Res.*, **11**, 66–77.
[76] Chan, P.C., Eustis, S.L., Huff, J.E. *et al.* (1988) *Toxicology*, **52**, 237–252.
[77] Husain, R., Khan, S., Husain, I. *et al.* (1989) *Ind. Health*, **27**, 121–124.
[78] Goldman, M., Dacre, J.C. (1989) *Rev. Environ. Contam. Toxicol.*, **110**, 75–115.
[79] Clark, D.N. (1989) *Review of Reactions of Chemical Agents in Water*, Defense Technical Information Center, Alexandria, VA.
[80] Epstein J., Callahan, J.J., Bauer, V.E. (1974) *Phosphorus*, **4**, 157–163.
[81] Kingery, A.F., Allen, H.E. (1995) *Toxicol. Environ. Chem.*, **47**, 155–184.
[82] Rosenblatt, D.H., Small, M.J., Kimmell, T.A. Anderson, A.W. (1995) *Agent Decontamination Chemistry: Technical Report*, Argonne National Laboratory, Washington, DC.
[83] Amr, A.T., Cain, T.C., Cleaves, D.J., *et al.* (1996) *Preliminary Risk Assessment of Alternative Technologies for Chemical Demilitarization*, Mitretek Systems, McLean,VA.
[84] National Research Council (1996) *Review and Evaluation of Alternative Chemical Disposal Technologies*, National Research Council, Washington, DC.
[85] D'Agostino, P.A., Provost, L.R., and Visentini, J. (1987) *J. Chromatogr.*, **402**, 221–232.
[86] Rohrbaugh, D.K. (1998) *J. Chromatogr.*, **809**, 131–139.

[87] Registry of Toxic Effects of Chemical Substances Database (1997) *Diisopropyl ethyl mercaptoamine*, MEDLARS Online Information Retrieval System, National Library of Medicine, Bethesda, MD, Document number: NTIS AD277-689.

[88] Szafraniec, L.J., Szafraniec, L.L, Beaudry, W.T., Ward, J.R. (1990) *On the Stoichiometry of Phosphonothiolate Ester Hydrolysis*, U.S. Army Chemical Research, Development and Engineering Center, Aberdeen Proving Ground, MD.

[89] Horton, R.J. (1962) *EA 2192: A Novel Anticholinesterase*, U.S. Army Chemical Research and Development Laboratories, Edgewood Arsenal, MD.

[90] Yang, Y.C., Szafraniec, L.L., Beaudry, W.T., Rohrbaugh, D.K. (1990) *J. Am. Chem. Soc.*, **112**, 6621–6627.

[91] MacEwen, J.D., Vernot, E.H. (1985) *Toxic Hazards Research Unit Annual Technical Report*, Harry G. Armstrong Aerospace Medical Research Laboratory, Wright-Patterson Air Force Base, OH.

[92] Smyth, H.F., Carpenter, C.P., Weil, C.S., Pozzani, U.C. (1954) *Arch. Ind. Hyg. Occ. Med.*, **10**, 61–68.

[93] Registry of Toxic Effects of Chemical Substances Database (1997) *2-Diisopropylaminoethanol. Union Carbide Data Sheet*, MEDLARS Online Information Retrieval System, National Library of Medicine, Bethesda, MD.

[94] Registry of Toxic Effects of Chemical Substances Database (1997) *Diethyl methyl phosphonate*, MEDLARS Online Information Retrieval System, National Library of Medicine, Bethesda, MD.

[95] Armstrong, D.J., Fukuto, T.R. (1984) *J. Agric. Food Chem.*, **32**, 774–777.

[96] Registry of Toxic Effects of Chemical Substances Database (1997) *O,S-Diethyl methylphosphonothioate*, MEDLARS Online Information Retrieval System, National Library of Medicine, Bethesda, MD.

[97] Toropkov, V.V. (1980) *Gig. Sanit.*, **45**, 79–81.

[98] Union Carbide Corporation (1971) *Diisopropylamine*, Union Carbide Corporation, New York.

[99] Izmerov, N.F., Sanotsky, I.V., Sidorov, K.K. (1982) Toxicometric Parameters of Industrial Toxic Chemicals under Single Exposure. United Nations Environment Programme, Paris.

[100] Kennedy, G.L.J., Graepel, G.J. (1991) *Toxicol. Lett.*, **56**, 317–326.

[101] Treon, J.F., Sigmon, H., Kitzmiller, K.V., Heyroth, F.F. (1949) *J. Ind. Hyg. Toxicol.*, **31**, 142–145.
[102] Final report on the safety assessment of diisopropylamine (1995) *J. Am. Coll. Toxicol.*, **14**, 182–195.
[103] Marhold, J.V. (1986) *Preheld Prumyslove Toxikologie: Organike Latky*, vol. 1, Avicenum, Prague, pp. 1–760.
[104] Company, M. (1985) *Nine Toxicity Studies on Diisopropylamine*, National Technical Information Service, Springfield, VA.
[105] Moyer, R.C. (1990) *Chem. Eng. News.*, **68**, 2.
[106] Eastman Kodak Company (1992) *Acute Toxicity Studies of N,N'-Methanetetraylbiscyclohexanamine*, Eastman Kodak Company, National Technical Information Service, Springfield, VA.
[107] Mortelmans, K., Haworth, S., Lawlor, T. *et al.* (1986) *Environ. Mutagen.*, **8**, 1–119.
[108] Gelernt, M.D., Herbert, V. (1982) *Nutr. Cancer*, **3**, 129–133.
[109] Microbiological Associates Inc. (1996) *Final Report of the Twenty-Week Prechronic Dermal Toxicity/Carcinogenicity Study of Diisopropylcarbodiimide (DIC) in Female NTPTG.AC Mice*, National Toxicology Program, Research Triangle Park, Document number: CAS No. 538-75-0. Study No. 93021B.
[110] Pathco Chairperson's Report (1997) *90-Day Subchronic Skin Painting Study of 1,3-Diisopropylcarbodiimide (DIC) in F344 Rats and B6C3F1 Mice (C93020-03/93020-04), and 150-day Subchronic Skin Painting Study of DIC in Female Transgenic TG.AC Mice (C93020-05)*, National Toxicology Program, Research Triangle Park, NC, Document number: N01-ES-15319.
[111] National Toxicology Program, *Diisopropylcarbodiimide. Salmonella Testing Results*, unpublished data.
[112] National Toxicology Program, *Diisopropylcarbodiimide. in vivo Cytogenetics Testing Results. Micronucleus Induction Results*, unpublished data.
[113] Immunotoxicology Program (1995) *Assessment of Contact Hypersensitivity to Diisopropylcarbodiimide in Female B6C3F1 Mice*, Virginia Commonwealth University, Richmond, VA.
[114] Poesen, N., de Moor, A., Busschots, A., Dooms-Goossens, A. (1995) *Contact Dermatitis*, **32**, 368–369.
[115] Microbiological Associates Inc. (1996) *Final Report of the Twenty-Week Prechronic Dermal Toxicity/Carcinogenicity Study*

of Dicyclohexylcarbodiimide (DCC) in Female NTPTG.AC Mice, National Toxicology Program, Research Triangle Park, NC.

[116] Pathco Chairperson's Report (1997) *Pathology Working Group (PWG) Review of N,N-Dicyclohexylcarbodiimide (DCC) 13-Week Subchronic Study in B6C3F1 Mice and F344 Rats, and 150 Day Subchronic Study in Female TG.AC Mice Administered by Skin Application*, National Toxicology Program, Research Triangle Park, NC, Document number: N01-ES-15319.

[117] National Toxicology Program, *Dicyclohexylcarbodiimide. Salmonella Testing Results*, unpublished data.

[118] National Toxicology Program, *Dicyclohexylcarbodiimide. Salmonella Testing Results*, unpublished data.

[119] Meade, B.J. (1996) *Assessment of Contact Hypersensitivity to Dicyclohexylcarbodiimide in B6C3F1 Female Mice*, Department of Pharmacology and Toxicology, Virginia Commonwealth University, Richmond, VA.

[120] William, D.M.S., Augerson, S. (2000) *A Review of the Scientific Literature as it Pertains to Gulf War Illnesses: Chemical and Biological Warfare Agents*, United States Department of Defense, Office of the Secretary of Defense, National Defense Research Institute (U.S.).

[121] Committee on the Survey of the Health Effects of Mustard Gas and Lewisite, Institute of Medicine (1993) *Veterans at Risk: The Health Effects of Mustard Gas and Lewisite*, (eds C.M. Pechura, and D.P. Rall), The National Academies Press, Washington, DC.

[122] Papirmeister, B., Feister, A.J., Robinson, S.I., Ford, R.D. (1991) *Medical Defense Against Mustard Gas*, CRC Press, Boca Raton, FL.

[123] Rosenblatt, D.H., Dacre, J.C., Muul, I., Cogley, D.R. (1975) *Physical, Chemical, Toxicological, and Biological Properties of 16 Substances. Tech. Rpt. 7509*, U.S. Army Medical Bioengineering Research and Development Laboratory, Fort Detrick, MD.

[124] Epstein, J. (1956) *Public Health Rep.*, 71, 955–962.

[125] NDRC (1943) *Office of Scientific Research and Development*, Volume 9-4-1-9, National Defense Research Committee, Washington, DC.

[126] M. O. I. R. System., National Library of Medicine, Bethesda, MD, 3 March 1998.

[127] Vilensky, J.A., Sinish, P.R. (2005) *Dew of Death: The Story of Lewisite, America's World War I Weapon of Mass Destruction*, Indiana University Press, Bloomington, IN.
[128] Minnesota Poison Control System, Hennepin County Medical Center (2004) http://www.mnpoison.org/index.asp?page ID= 146 (accessed 5 June 2010).
[129] Walton, L.P., Murray, V.S.G. (1996) *ICPS InChem.*, http://www.inchem.org/documents/pims/chemical/mustardg.htm (accessed 13 May 2010).
[130] Reddy G., Major, M.A., Leach, G.J. (2005) *Int. J. Toxicol.*, **24**, 435–442.
[131] Agency for Toxic Substances and Disease Registry (2009) *Medical Management Guidelines for Blister Agents: Nitrogen Mustard (HN-1) ($C_6H_{13}Cl_2N$), Nitrogen Mustard (HN-2) ($C_5H_{11}Cl_2N$), Nitrogen Mustard (HN-3) ($C_6H_{12}Cl_3N$)*, http://www.atsdr.cdc.gov/mhmi/mmg164.html (accessed 13 June 2010).
[132] Yongsheng, D., Kim, R. (2008) *Electroanalysis*, **20**, 2192–2198.
[133] Isachi, O., Koichiro, T., Hisashi, M. *et al.* (2005) *Jpn. J. Forensic Toxicol.*, **23**, 118–119.
[134] Holstege, C.P. (2008) *Vomiting Agents-Dm, Da, Dc*, http://emedicine.medscape.com/article/833391-print (accessed 29 July 2009).
[135] Federation of American Scientist (1996) *Riot Control Agents*, http://www.fas.org/nuke/guide/usa/doctrine/dod/fm8-9/3ch7.htm (accessed 29 July 2009).
[136] Henriksson, J. (1996) *Arch. Environ. Contam. Toxicol.*, **30**, 213–219.
[137] Ochi, T. (2004) *Toxicol. Appl. Pharmacol.*, **200**, 64–72.
[138] Kroening, K.K. (2009) *Metallomics*, **1**, 59–66.
[139] Kazuhiro, I. (2004) *Ann. Neurol.*, **56**, 741–745.
[140] Kato, K. (2007) *Life Sci.*, **81**, 1518–1525.
[141] Sanderson, H. (2008) *J. Hazard. Mater.*, **154**, 846–857.
[142] Sanderson, H. (2009) *J. Hazard. Mater.*, **162**, 416–422.

3

Analysis of Chemical Warfare Agents

Analysis of Chemical Warfare Degradation Products, First Edition. Karolin K. Kroening,
Renee N. Easter, Douglas D. Richardson, Stuart A. Willison and Joseph A. Caruso.
© 2011 John Wiley & Sons, Ltd. Published 2011 by John Wiley & Sons, Ltd.

3.1 INTRODUCTION

The analysis of chemical warfare agents is dependent on a myriad of factors, such as sensitivity, selectivity (especially in environmental matrices and complex mixtures), reproducibility, quick analysis and throughput, and analysis with little to no sample preparation. Ideally, one instrument would satisfy all of the requirements necessary for a fast and positive identification. With the advancement of technology, this is slowly becoming a reality. Currently, there are many different approaches worth pursuing for the analysis of the aforementioned nerve agents, mustard agents and sternutator agents. The most widely used methods to date for verification and quantification of nerve agents are chromatographic techniques, specifically gas chromatography (GC) or liquid chromatography (LC), typically coupled with mass spectrometry (MS) or tandem MS, (GC-MS/MS) or (LC-MS/MS).

There are more approaches to the analysis of chemical warfare agents than just the chromatographic or mass spectrometric based techniques. These include nuclear magnetic resonance spectroscopy (NMR) or spectrophotometric detection. Additional approaches similar to GC, LC and MS are capillary electrophoresis (CE) and ion mobility spectroscopy (IMS). This chapter describes the instrumentation used to analyse and detect chemical warfare agents, from the simplest of techniques to the more complex, which allow for precise and accurate detection at very low levels. The authors have tried to be as comprehensive as possible in outlining the various methods utilized; however, it is understood that because of the implications of CWAs, there may be proprietary or classified methods, not yet in the public domain.

3.2 MINIMALLY INVASIVE DETECTION TECHNIQUES

Some of the simplest approaches used to detect chemical warfare agents usually incorporate minimal or non-invasive techniques. These include acoustic analysis, neutron activation analysis, nuclear magnetic resonance (NMR) spectroscopy, spectrophotometric detection, electrochemical detection and immunochemical detection. In order to obtain acceptable results for the accurate determination of nerve agents, mustard agents and sternutator agents, these techniques must be useful specifically on neat or relatively pure agents, where any minimal interferences are negligible. Otherwise, they have to be used in conjunction with another technique or implemented for screening method purposes only.

Acoustic analysis detects changes in the properties of acoustic waves as they travel at ultrasonic frequencies in piezoelectric materials. The interaction between the waves and the phase-matter composition facilitates chemical selectivity and, thus, the detection of CWA's. These are commonly known as surface acoustic wave (SAW) sensors. Reported studies indicate detection limits as low as $0.01\,\mathrm{mg}\ \mathrm{m}^{-3}$ for organophosphorus analytes within a 2 min analysis [1]. There are several commercially available SAW instruments, which can automatically monitor for trace levels of toxic vapors from G-nerve agents and other CWAs, with a high degree of selectivity. A major advantage of SAW detectors is that they can be made small, portable and provide a real-time analysis of unknown samples. One of the drawbacks of these instruments is that sensitivity and a rapid response time are inversely related. In an ideal instrument, both parameters would be obtained without sacrificing one for the other.

Another disadvantage is that selective coatings are necessary in order to identify specific chemical analytes. The production of selective polymer films is difficult and these polymer films are susceptible to surface changes, known as dewetting, which leads to a decreased response and instrument performance [2]. As a result, false positives and selectivity issues can occur [1].

The working principle of how a binary channel surface acoustic wave (SAW) lithium niobate piezoelectricn chip detects Sulfur Mustard and established the mathematical model of beat (MEMS) craft led to the solution for the parameters of the binary channel SAW chip. According to thermogravmetric and differential scanning calorimetry (TG-DSC) analysis, vacuum coating craft was adopted to solve the hybrid sensitive film forming craft parameter of $PdPc_{0.3}PANI_{0.7}$ (phthalocyanine palladium 0.3–polyaniline 0.7). The micro-appearance of the sensitive film was analysed using scanning electron microscopy (SEM) [3].

Neutron activation analysis irradiates stable, naturally occurring elements with neutrons, promoting nuclei to excited states by neutron capture and neutron inelastic scattering reactions. As these nuclei relax, they emit characteristic gamma rays, which can be measured and compared with the gamma ray spectrum, allowing for identification of the chemical elements present. The resulting analysis of the stoichiometric ratios of these elements allows for the determination of the chemical composition. Although chemical composition is not a substantial amount of information to be gained in the determination of chemical warfare agents, given the limited number of chemical types found in munitions and weapons, information on which elements are present and which are absent is often sufficient to accurately infer their identity. As nerve agents contain phosphorus, an indication of a high phosphorus concentration in the elemental composition could indicate the presence of phosphorus-containing nerve agent or similar compounds. Distinguishing between VX and the G-series of nerve agents is possible by establishing the positive presence of sulfur. In the same way, the determination of Sulfur Mustard would be relevant if a high sulfur concentration was detected. The blister agents Mustard and Lewisite contain phosphorus, but both contain about 45–50 weight-% chlorine, and Lewisite can be distinguished from the Mustard agent by the absence of sulfur and the presence of arsenic [4]. This type of technique is used for sealed munitions in particular.

A neutron diagnostic experimental apparatus has been designed for nondestructive verification of preserved weapons. This equipment is based on an easy to use industrial sealed tube neutron generator that interrogates the munitions of interest with 14 MeV neutrons. Gamma-ray spectra are detected with a high purity germanium detector, while shielded from neutrons and gamma ray background. Possible configurations allow the detection of gamma rays from neutron inelastic scattering, (in continuous or pulsed modes) from thermal neutron capture and fast or thermal neutron activation. Tests on full scale sealed munitions with chemical simulants show that those with chlorine (old generation materials) are detectable in a few minutes, and those including phosphorus (new generation materials) take almost the same time [5].

Prompt gamma ray neutron activation analysis (PGNAA) is a proven method for the identification of chemical warfare agents, which is widely used as a nondestructive technique.

An interesting application can be found in the literature, from Anikiev and Kolesov [6]. They described an analysis of dumped chemical warfare agents in marine bottom sediments. The method determins the characteristic ratios between primary elements of the marine sediment and

Sulfur Mustard (oxygen, carbon, nitrogen, phosphorus, sulfur) by electron spectroscopy, and determining the fractionation coefficient for arsenic incorporated into Lewisite relative to scandium using neutron activation analysis. Scandium is used for the estimation of the effects of geochemical factors on the chemical composition of the marine sediment was also described.

NMR is an important technique for the structural characterization of chemical warfare agents, yet its value for complex mixtures at trace amounts is still limited. ^{1}H, ^{13}C and ^{19}F NMR can be selected to identify nerve agents and their breakdown products. The most useful NMR technique would be to analyse target nerve agents using ^{31}P NMR, because of the characteristic chemical shifts of compounds containing a phosphorus–carbon bond [2–5]. ^{31}P chemical shifts are sensitive to temperature, concentration and solvent, all of which can hinder a rapid identification. This technique, although potentially valuable, should only be used to screen for higher levels of nerve agents in conjunction with other methods.

The use of NMR related to mustard agents has been reported by Arroyo and coworkers [7]. The aim of this study was to apply the magnetic resonance technique and electron paramagnetic resonance (EPR), to obtain a complete picture of the mechanism of the chemical reaction of a selected polyoxometalate (POM) with Sulfur Mustard and its analogous compounds. This is because vanadium-containing polyoxometalates of Keggin structure $H_5PV^V_2Mo_{10}O_{40}$ can activate sulfides and oxidize such substrates, yielding the product and a reduced catalyst. Reoxidation of the catalyst by dioxygen and formation of water then follows. The regeneration of the original oxidized form of the catalyst by reaction with a molecule of oxygen is very fast suggesting that $H_5PV^V_2Mo_{10}O_{40}$ could be used as a catalyst for the decontamination of Sulfur Mustard and its analogs [7].

Clark I, II and Adamsite can also be detected by NMR techniques. Using ^1H and ^{13}C along with CH-COSY (correlation spectroscopy) and NOESY (nuclear Overhauser effect spectroscopy) experiments. Mesilaakso and coworkers were able to obtain spectra of all three analytes in three different solvents, $CDCl_3$, CD_2Cl_2 and $(CD_3)_2CO$ [8]. They also tested a range of analysis temperatures. Between the solvents and the temperatures, they were able to show that overall the same spectra were obtained and chemical shifts can easily be determined when experimental parameters vary.

Electrochemical determination of organophosphate nerve agents can also be achieved using an enzyme biosensor [6, 7]. The biosensor is integrated with the enzyme, organophosphorus hydrolase, on glass beads. The enzyme hydrolyzes organophosphate to an electroactive species, which can be determined amperometrically. A single-channel microchip was fabricated using this approach for the rapid screening of organophosphate nerve agents [9]. Chemical resistant vapor detectors have also been reported [10].

Electrochemical detectors are available throughout the world for the nerve, choking and blood agents. However, electrochemical detectors are not available for the detection of blister agents due to their non-electrochemical activity on the electrode surface [9].

Electrode material and surface state have a considerable effect on the electrochemical reaction that could detect Sulfur Mustard; hence, modification of the electrode surface is the only way to obtain the electrochemical activity from blister agents. By selecting a carbon material, selectivity and sensitivity of the gas are controlled using its surface properties. A compact type of electrochemical sensor to detect blister agents, such as Sulfur Mustard or Lewisite 1, using a carbon electrode modified with gold nanoparticles has been studied. A detection level of 1.5 mg m^{-3} was observed [11]. Another study reported the electrochemical preparation of a composite film on a gold electrode surface consisting of polypyrrole and CuPc with the cationic surfactant cetyltrimethylammonium bromide (CTAB). Combined properties of metal phthalocyanine and a conducting polymer enhanced the detection of Nitrogen Mustard-1. An FTIR (Fourier transform infra-red) study was conducted to find out the extent of the immobilization of CuPc with the polymer. The morphologies and elemental composition of the pPy/CuPc/CTAB modified electrode were obtained from SEM and EDS (energy dispersive X-ray spectroscopy) analyses, respectively. Finally, the modified electrode was characterized electrochemically and utilized for the sensing of blistering chemical warfare agent Nitrogen Mustard-1 [8]. Berger and coworkers have employed an Au-microelectrode sensor to develop a potentiodynamic sensor system for the detection of Clark I and other agents [12]. The sensor reduces and oxidizes each sample differently allowing for identification of each analyte through various voltammogramms.

Colorimetric responses can be obtained using detection paper, which relies on certain dyes being soluble in chemical warfare agents. Normally, two dyes and one pH indicator are used, which are mixed with cellulose fibers in a paper without any special coloring (unbleached). When a drop of a chemical warfare agent is absorbed by the paper, it dissolves one of the pigments. Mustard agent results in a red color change while the presence of a nerve agent results in a yellow color change. The presence of VX gives a different color change from other nerve agents, turning the indicator paper to a green/green–black color.

Detection paper can thus be used to distinguish between three different types of chemical warfare agents. A disadvantage with the papers is that many other substances can also dissolve the pigments. Consequently, they should not be located in places where drops of solvent, fat, oil or fuel can fall on them. Drops of water do not trigger a reaction.

On the basis of spot diameter and density on the detection paper, it is possible estimate on the original size of the droplets and the degree of contamination. A droplet of 0.5 mm diameter gives a spot sized about 3 mm on the paper. A droplet/cm^2 of this type corresponds to a ground contamination of about 0.5 g m^{-2}. The lower detection limit in favorable cases is 0.005 g m^{-2} [10].

M8 and M9 are common chemical agent detection papers, used by the military, but are also available commercially to hazardous material (HAZMAT) response teams [10]. In the case of M8 paper (2.5 × 4 in), it is blotted onto a suspected liquid agent and observed for a color change within 30 s. V-type nerve agents turn the M8 paper dark green, G-type nerve agents turn it yellow and blister agents present a red color. The M9 detection paper is more of an adhesive type patch to be worn in the field, or it can be attached to vehicles that are entering areas filled with vapor to determine contamination. When it comes in contact with a liquid CWA droplet, the paper turns from the original green to red or pink. Other examples of kits are M256/M256A1, which can detect nerve gas concentrations down to 0.005 mg m^{-3} and Sulfur Mustard down to concentrations of 0.02 mg m^{-3}, in addition to M18 [13]. There are also many other colorimetric kits, tubes, chips and types of papers capable of detecting chemical warfare agents listed in the Responder Knowledge Base as well as in First Responder Guides [13–15]. These kits should only be used as an initial screening method due to their propensity for false positives and lack of sensitivity and selectivity for CWA's [14].

Spectrophotometers are more sensitive and reliable for the detection of chemical warfare agents than paper detection kits. These include identification by an ultraviolet (UV) or infra-red (IR) detectors. Although spectrophotometers are more sensitive detectors, one major problem in the case of nerve agents is that alkylphosphonic acids do not have chromophores or fluorophores. Detection by these methods relies on host chromophores or fluorophores to which the nerve agents will bind [16]. Techniques involving UV detection incorporate a lanthanide complex that will chelate to the nerve agent resulting in luminescence [17, 18]. An opposing approach incorporates the quenching of the luminescence taking place upon exposure to the nerve agent [19]. One of the difficulties with using lanthanide complexes is their propensity to quench in the presence of water, which would result in a false detection in many environmental matrices [20]. IR spectroscopy offers information about the different functional groups present, but not about the molecule as a whole. Therefore, IR methods can be used as a screening process in order to provide structural details of the molecules present or as a complementary tool to support other analysis techniques. IR spectroscopy has been used to detect nerve agents such as Sarin in water [21]. Typically IR devices are paired with another analysis technique such as GC or GC-MS [22–24].

3.3 SEPARATION AND DETECTION TECHNIQUES

Frequently used screening and quantitative methods for iden-tification of nerve agents, blister agents and sternutator agents involve the use of GC, LC and CE. Several useful reviews regarding the characterization and quantification of CWA's using chromatographic and mass spectrometric techniques have been conducted [25–29].

3.3.1 Capillary Electrophoresis

Capillary electrophoresis (CE) is a separation technique for ionic or ionizable compounds. CE is particularly attractive because the instrumentation is inexpensive and separations are quick and efficient. As with GC and LC, CE can be coupled to and flame photometric detection (FPD) to detect alkylphosphonic acids [30–32]. Indirect UV absorbance detection with CE has also been used for the analysis of nerve agents and their degradation products [33]. In an attempt to meet the demands of portable and efficient field instruments, miniaturized analytical systems with CE microchips have also been made for the separation and detection of alkylphosphonic nerve agents [34]. The afore-mentioned CE procedures all provide rapid identification without extensive sample preparation. CE is most likely to be used as a guide in order to select the appropriate methods for further analysis by more definitive techniques such as GC-MS, as most of the products detected and analysed are degradation products [35]. A review depicting various CE separation techniques, lab-on-a-chip technology and detection limits has been compiled by Pumera and is shown in Table 3.1.

Capillary electrophoresis for the determination of Sulfur Mustard, sternutator agents or associated compounds is not used to any great extent because the analytes are charge deprived [15].

Table 3.1 Analysis of nerve agent degradation products using capillary electrophoresis. Reprinted from Journal of Chromatography A, 1217, 45, Copyright 2010, with permission from Elsevier

Degradation analyte	Sample matrix	CE separation technique parameters	Detection technique	Detection limit	Analysis time (min)	Reference
EMPA, IMPA, PMPA, MPA	–	–	Direct UV at 210 nm	300 µg L^{-1}	7	16
EMPA, IMPA, PMPA, MPA	Groundwater	CE with EOF reversor	Indirect UV at 254 nm (4.5 mM chromate ion as ultraviolet visualization agent)	Low mg L^{-1} levels	5	17
EMPTA, BMPTA	Environmental water	MEKC (100 mM SDS)	UV at 200 nm	1–10 mg L^{-1}	10	18
EMPA, IMPA, PMPA, CMPA, MPA	Soils	CE with EOF suppressor (decamethonium bromide)	Indirect UV at 254 nm (10 mM sorbate ion as ultraviolet visualization agent)	~5 mg L^{-1}	15	19

Analytes	Matrix	Method	Detection	LOD		
EMPA, IMPA, PMPA, CMPA, MPA	–	CE with EOF reversor (polybrene)	Indirect UV at 254 nm (5 mM sorbate ion as ultraviolet visualization agent)	–	6	20
EMPA, EMPTA, IMPA, PMPA, CMPA, MPA	Surface water, soils, paint, rubber	CE with EOF reversor	Flame photometric detection	Low mg L^{-1} levels	20	21
EMPA, IMPA, PMPA, CMPA, BMPA, MPA, IEPA, EHMPA	Soils, surfaces, vegetation	CE with EOF reversor (didodecyl-dimethyl-ammonium hydroxide or cetyltrimethyl-ammonium hydroxide)	Conductivity and indirect UV detection at 210 nm (10 mM phenylphosphonic acid as ultraviolet visualization agent)	75 µg L^{-1} (conductivity detector); 100 µg L^{-1} indirect UV detector	3	22

Table 3.1 (*continued*)

Degradation analyte	Sample matrix	CE separation technique parameters	Detection technique	Detection limit	Analysis time (min)	Reference
IMPA, GB-MEA adduct, PMPA, GD-MEA adduct	—	CE with EOF reversor (didodecyl-dimethyl-ammonium hydroxide)	Conductivity and indirect UV detection at 210 nm (10 mM phenylphosphonic acid as ultraviolet visualization agent)	100 μg L^{-1}	7	23
EMPA, IMPA, PMPA, MPA	Surface water, groundwater, soils	CE with EOF reversor (tetradecyl-trimethyl-ammonium hydroxide)	Conductivity detector	6–60 mg L^{-1}	10	24
EMPA, IMPA, PMPA, CMPA, CPMPA, MPA	Tap water	—	Simultaneous indirect UV at 254 nm (5 mM sorbic acid as ultraviolet visualization agent) and MS detection; MS–MS detection	5 mg L^{-1} for MS detection, 100 μg L^{-1} for MS–MS	25	25

Analytes	Matrix	Method	Detection	LOD		Ref
EMPA, IMPA, PMPA, MPA	Environmental water and soils	CE with EOF reversor (didodecyl-dimethyl-ammonium hydroxide)	Conductivity and indirect UV detection at 210 nm (10 mM phenylphosphonic acid as ultraviolet visualization agent)	1–2 µg L^{-1}	3	26
EMPA, IMPA, PMPA, CMPA, BMPA, MPA	Human serum	CE with EOF reversor (didodecyl-dimethyl-ammonium hydroxide)	Indirect UV detection at 210 nm (10 mM phenylphosphonic acid as ultraviolet visualization agent)	100 µg L^{-1}	14	27
EMPA, IMPA, PMPA, MPA	–	–	Indirect LIF detection (8-hydroxypyrene-1,3,6-trisulfonic acid as the indirect probe)	40–80 µg L^{-1}	2	28
EMPA, EMPTA, IMPA, PMPA, MPA	Environmental water, soils	–	Flame photometric detection	1 mg L^{-1}	16	29

Table 3.1 *(continued)*

Degradation analyte	Sample matrix	CE separation technique parameters	Detection technique	Detection limit	Analysis time (min)	Reference
EMPA, IMPA, PMPA, MPA	—	CE with EOF reversor (didodecyl-dimethyl-ammonium hydroxide)	Indirect UV detection at 214 nm (1 mM phenylphosphonic acid as ultraviolet visualization agent)	100 nM	5	30
MPA	—	—	LIF	130 nM	35	31
2-(Dimethylamino) ethanethiol, 2-(diethylamino) ethanethiol, and cyanide	—	—	LIF	2–35 mg L^{-1}	6	32

3.3.2 Ion Mobility Spectrometry

Ion mobility spectrometry (IMS) operates by separating ions according to their size-to-charge ratio. Ions pass through a gas in an electric field and the differences in velocities result in separation. Under low electric field conditions, the ion mobility is directly proportional to the electric field strength. High electric field strength ion mobility instruments are becoming available, but lack sensitivity [36]. Other types of instrumentation include aspiration IMS detectors, which involve ions traveling through an orthogonal electric field, separated through multichannels and collected onto an electrode for the detection of the nerve agents [37]. These too suffer from low resolving power, increasing the possibility for false positives. Typical portable IMS units, such as the SABRE 4000, are capable of detecting Sarin, Soman, Cyclohexylsarin, Tabun and Nitrogen Mustard-3, each with a limit of alarm (LOA) of $0.005-0.5$ mg m^{-3} in positive ion collection mode. Hydrogen cyanide could be identified with an LOA of 0.2 mg m^{-3} in the negative mode. Sulfur Mustard, Nitrogen Mustards-1, -2 and -3, phosgene and chloropicrin showed a positive alarm of 'HD/Phos' with an LOA of $0.2-2$ mg m^{-3} in the negative mode. IMS can be coupled to mass spectrometers and gas chromatography devices for the reduction of false positives and greater sensitivity, yet are then no longer portable for field evaluations (Table 3.2) [38].

Most IMS instruments are portable and light-weight, and for screening purposes are designed to detect chemical warfare agents in the field [39–40]. IMS is the cornerstone of many devices used today (Figure 3.1). The Finnish M86 and M90 are handheld devices that use IMS, as is the Improved Chemical Agent Monitor (ICAM). The ICAM was used extensively in the Gulf War, and was even attached to certain vehicles. It is a handheld device, which continuously

Table 3.2 Selected IMS detectors and their assessment criteria. Reprinted from Journal of Chromatography A, 1217, 45, Copyright 2010, with permission from Elsevier

Criteria	CAM [33–35]	APD 2000 [36]	Multi-IMS [37]	Sabre 4000 [38, 39]
Detected agents	Blood, blister, choking and nerve agents plus selected TICs	GA, GB, GD, VX, HD, L, pepper spray and mace	Nerve, blister, blood and choking agents	GA, GB, GD, GF, VX, vesicants, TICs, drugs and explosives
Limits of detection	LODs in line with or exceed the NATO requirements	V agents – 4 ppb; G agents – 15 ppb; H – 300 ppb; L – 200 ppb	Nerve – 0.01 –0.1 mg m^{-3}; Blister – 0.5 –2.0 mg m^{-3}; Blood/choking – 20 –50 mg m^{-3}	–
Simultaneous detection	No	Yes for nerve and blister agents. To detect irritants the mode must be manually changed	Yes	No
Operational temperature range (°C)	–25 to +55	–30 to +52	–30 to +50	0 to +45

Figure 3.1 Chemical detection equipment. Fixed Site/Remote Chemical Agent Detector using ion mobility spectroscopy (IMS). Image courtesy of Environmental Technologies Group Inc., Hauppauge, NY

displays the concentration of nerve or mustard agents. The ICAM is prone to erroneous detection in enclosed spaces and areas of strong vapor concentration (e.g., heavy smoke). It can also become saturated, requiring recalibration. Versions of the ICAM are available for commercial purchase and are used by many medical response teams [40].

Usually detection limits for sulfur mustard are in the region of $0.1\,mg\;m^{-3}$ [41–46]. An interesting application of a miniaturized aspiration condenser-type ion mobility spectrometer for fast detection of chemical warfare agents has been reported. The device was tested at the Armed Forces Scientific Institute for Protection Technologies-NBC-Protection, Germany, to evaluate the analytical performance. The spectra of different chemical warfare agents, such as Sarin, Tabun, Soman, US-VX, Sulfur Mustard, Nitrogen Mustard and Lewisite were recorded at various

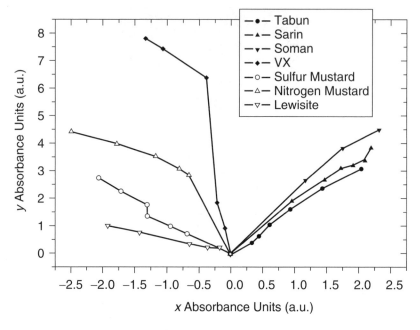

Figure 3.2   The patterns of Tabun, Sarin and Soman are very similar, whereas the patterns of US-VX, Sulfur Mustard, Nitrogen Mustard and Lewisite are very different. Reprinted from Stefan Zimmermann, Sebastian Barth, Wolfgang K.M. Baether and Joachim Ringer. Miniaturized Low-Cost Ion Mobility Spectrometer for Fast Detection of Chemical Warfare Agents. *Anal. Chem.*, 2008, 80 (17), pp. 6671–6676. Copyright © 2008 American Chemical Society

concentrations (Figure 3.2). From earlier tests it was known that lower concentrations of Sulfur Mustard, Nitrogen Mustard and Lewisite are not detectable under humid conditions. Thus, purified dry air with an absolute humidity of 20 ppm was used for all experiments to evaluate the separation power of the IMS concept, independent of any inlet system. On the basis of these measurements, it was demonstrated that the separation power of this miniaturized ion-focusing aspiration condenser IMS is sufficient to clearly identify all trained substances down to a few micrograms

Table 3.3 Identification limits C_{id} and lowest concentrations trained C_{min}. Australian Government Department of Defence. Defence Science & Technology Organisation - DSTO-GD-0570

Chemical warfare agent	C_{id} ($\mu g\ m^{-3}$)	C_{min} ($\mu g\ m^{-3}$)
Tabun	2	3
Sarin	6	15
Soman	7	30
US-VX	3	30
Sulfur Mustard	34	70
Nitrogen Mustard	24	150
Lewisite	93	63

per cubic meter (Table 3.3). However, the identification algorithm used does not yet allow substance identification in the presence of ionized interferents [47].

Recent literature has reported on the testing of a portable ^{241}Am ionization aspiration-type ion mobility spectrometer (M90-D1-C, Environics Oy), and it was investigated with nerve gases, blister agents, and blood agents. The vapors of the nerve gases, Sarin, Soman, Tabun and Cyclohexylsarin, were recognized as 'NERVE' after several seconds of sampling, and the limits of detection (LOD) were <0.3 mg m^{-3}. The vapors of blister agents, Sulfur Mustard and Lewisite 1, and blood agents, hydrogen cyanide and cyanogen chloride, were recognized as 'BLISTER' with an LOD of <2.4 and >415 mg m^{-3}, respectively. The vapors of Nitrogen Mustard-1 and -2 and phosgene did not show any alarm [48].

3.3.3 Gas Chromatography (GC)/Gas Chromatography-Mass Spectrometry (GC-MS)

Gas chromatography is one of the most universally implemented chromatographic techniques for volatile compounds because of its sensitivity, selectivity and ease of use. Because nerve agents are thermally stable, element-selective detectors, such as flame photometric detection (FPD) or nitrogen–phosphorus detection (NPD) used with GC analysis, allow for fast detection in different matrices and mixtures. Studies have used GC-FPD and GC-NPD to test and improve extraction procedures because of the enhanced selectivity achieved for nerve agents with detection limits as low as 0.05 ppb in water samples [41–44].

GC-atomic emission detection (AED) is yet another selective detector that has been used for the analysis and identification of CWA nerve agents [45–47]. One of the disadvantages of GC-AED is that in samples with a high hydrocarbon content, quenching of the plasma occurs during analysis [49]. Pulsed FPD (PFPD) works by separating the emission of carbon from phosphorus in time, resulting in an even more sensitive and selective approach with detection limits as low as 7 fg s^{-1} for phosphorus [48, 49]. An alternative approach incorporates a two-dimensional GC-GC analysis for highly complex environmental samples [50]. This allows for greater separation and chromatographic resolution of nerve agents from potential interferences.

Other examples of developments in GC are large-volume injections, for greater identification of nerve agents, high-speed GC, comprehensive GC-GC, or programmed temperature vaporizing (PTV) solvent split injection. Since their inception in the late 1960s and early 1970s, these techniques have become a viable avenue to use for detection, although today, most GC instruments are coupled to a MS detector for qualitative and quantitative purposes.

GC-MS detection devices are more commonly used today for the detection of nerve agents [51, 52]. GC-MS under electron impact (EI) conditions results in extensive fragmentation, providing structural information. GC-MS under chemical ionization (CI) conditions, which is a softer technique, reduces fragmentation, providing molecular mass information and a more sensitive detection of the targeted analytes. Several studies have been carried out using both EI and CI techniques coupled to GC-MS for the identification of nerve agents and their degradation products [53–58]. A majority of these detectors are designed to be used specifically for environmental samples; however, very few methods using GC-MS have been designed for detection of nerve agents in humans who have been exposed. As a consequence of the terrorist attacks that occurred within the Tokyo, Japan, subway system on 20 March, 1995, several methods have been developed using GC-MS for the identification of nerve agents in biological matrices [59–62]. Today a broader identification of nerve agents is possible in both environmental and biological matrices.

For enhanced sensitivity and lower quantitation levels, GC-tandem mass spectrometry (MS/MS) has also been used in several studies for the analysis of nerve agents [63–66]. Low levels of nerve agents have been detected in diesel exhaust using this technique. GC-MS/MS is essential for complex environmental samples where interferences could lead to either false positives or co-eluting compounds that complicate identification, resulting in a missed detection.

Today, with advanced technology, GC-MS instrumentation is now capable of being portable for use in the field for real-time analysis of agents within environmental samples [54, 67]. GC is considered one of the most prominent avenues for detecting CWA nerve agents to date on-going efforts to couple detection devices to GC instrumentation is desired for greater selectivity and sensitivity to confirm the presence of nerve agents [68].

GC is also the most widely used technique for the detection of Sulfur Mustard due to the volatility of the compounds. FPD and AED coupled to GC are frequently used detection methods [50–53].

A recent study published in the *Chinese Journal of Instrumental Analysis, Fenxi Ceshi Xuebao*, showed a detection limit of $500 \, \text{ng g}^{-1}$ of Sulfur Mustard (HD) by using accelerated solvent extraction–gas chromatography (ASE-GC) coupled with a flame photometric detector (FPD) in the sulfur mode, in soil. In this case, the study showed evidence that ASE results in better recoveries and sensitivity than liquid solid extraction (LSE) [50]. In 1996, a paper was published on a method for the analysis of Lewisite through derivatization of the compound before introduction into a gas chromatograph. In order to simplify the derivatization process, a tube packed with absorbent was used for collection of airborne vapors. If a positive response occurs when screening analytes using a GC coupled with an FPD, then the same sample can be analysed using a GC equipped with an AED for confirmation based on the elemental response to arsenic (in the case of Lewisite) and sulfur (in the case of Sulfur Mustard) within the appropriate GC retention time window [54].

Determining sulfur by FPD can represent a problem, as the sulfur-selective detector response is not linear and suffers from quenching by co-eluting hydrocarbons. An alternative to using this detector would be the use of sulfur chemiluminescence detection (SCD). SCD is based on forming sulfur monoxide in a reducing flame from the sulfur-containing compounds being eluted from a gas chromatograph. The sulfur monoxide that is formed is subsequently detected by its chemiluminescent reaction with ozone. This detector was found by these workers to be at least one order of magnitude more sensitive than most flame photometric detectors. It produces a linear response, possesses a sulfur-to-carbon selectivity greater than 106, does not suffer appreciably from quenching and has a response to various sulfur species that is almost equimolar [15, 55].

As mentioned above, AED is a technique particularly well-suited to screen complex samples for multiple compounds containing heteroatoms, such as phosphorus, sulfur or nitrogen, which are especially relevant in the study of chemical warfare agents. Among other GC detectors, AED has unique characteristics, such as compound independent calibration and possible raw formula determination [56]. Albro and Lippert have summarized the strengths and weaknesses of the performance of AED relative to other classical detectors for the analysis of chemical warfare agents and their breakdown products. Their conclusions illustrated that AED offers the best combination of sensitivity and selectivity of the detectors investigated for the analysis of sulfur compounds [57].

A curious and unfortunately sad study was carried out by Mazurek *et al.* [53]. Their work was on analysis to identify Sulfur Mustard in an yperite block, caught in the fishing nets of the crew of the cutter WLÇA 206 in the Baltic Sea in 1997. As a result of taking this unknown block of yperite on board, eight fishermen were poisoned and several others hospitalized. Ten random samples were taken from the block and extraction with 50 mL of dichloromethane. The samples were analysed by the military service using GC-AED and GC-MS. Although they set out to determine the concentration of yperite and its degradation products, Mazurek *et al.* were able to identify Clark I 19.76 min, and confirm by monitoring for arsenic at wavelength 189 nm. The results obtained for both Clark I and yperite show that after 50 years, CWAs can still be found in their original form (Figure 3.3) [53].

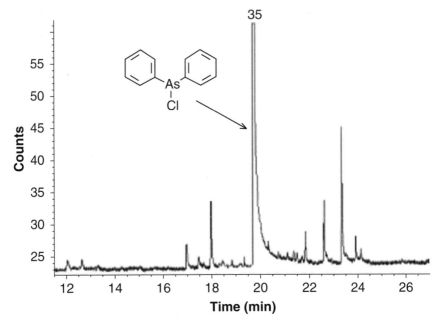

Figure 3.3 Dectection of Clark I by GC-AED by As 189 nm [53]. Reprinted from Mazurek, M., Witkiewicz, Z., Popiel, S., Sliwakowski, M. Capillary gas chromatography–atomic emission spectroscopy–mass spectrometry analysis of sulphur mustard and transformation products in a block recovered from the Baltic Sea, *Journal of Chromatography A* 2001, 919, 133–145 with permission from Elsevier

The peak identification of yperite was carried out by a comparison of the retention time of its standard with the retention times of the appropriate compound of the samples. Bis(2-chloroethyl) sulfide standard in dichloromethane and the solutions of mixtures containing the investigated samples were used for the comparison. It was found that the largest peak, with a retention time of 12.03 min (peak 17 in Figure 3.4), corresponds to yperite [53].

Gas chromatography-mass spectrometry (GC-MS) is currently the 'workhorse' of detection technologies, but carries a

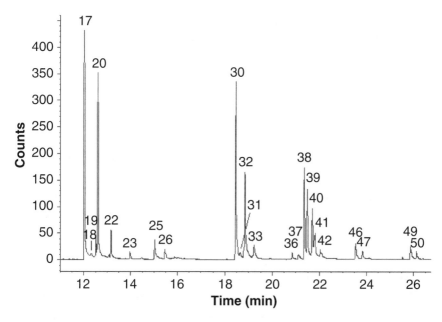

Figure 3.4   Element chromatogram of a sample of the yperite block obtained by GC-AED on sulfur, S-181 nm [53]. Reprinted from Mazurek, M., Witkiewicz, Z., Popiel, S., Sliwakowski, M. Capillary gas chromatography–atomic emission spectroscopy–mass spectrometry analysis of sulphur mustard and transformation products in a block recovered from the Baltic Sea, *Journal of Chromatography A* 2001, 919, 133–145 with permission from Elsevier

correspondingly high price tag. The sensitivity and selectivity of GC-MS are offset by the complexity of the instrumentation and lack of portability [49, 58–60].

A GC-MS-selected ion monitoring (SIM) method for the simultaneous detection of Sarin, Soman, Tabun, GF, VX, Russian VX and Sulfur Mustard in water was reported in 1999 [62]. Water samples were extracted twice with dichloromethane. The combined extracts were concentrated to 1 mL under a gentle stream of nitrogen and an internal standard solution was added. Selected ion monitoring for gas chromatography-mass spectrometry was used and the reported absorbance recovery rates were within the range of 82.6–93.3%, except for Sulfur Mustard with 67.8%. The lowest detection limits were $0.02-0.06 \, mg \, L^{-1}$, which were lower than the safe concentration for drinking water under field conditions [62].

Sulfur Mustard stability in nonpolar solvents has been determined by GC and GC-MS methods. The situation is more complex for Lewisite because GC methods generally involve derivatization with thiols, and are also complicated by the fact that Lewisite and its hydrolysis products give the same compound after derivatization. The solution to this problem can be found by measuring Lewisite without derivatization. In a study reported by Down in 2005 [63], toluene was selected as the extraction solvent because Lewisite slowly decomposed in other organic solvents, such as acetone and hexane. Thermal oligomerization in the injection port was prevented by on-column injection and a deactivated guard column was used to prevent the well-known problems of memory effects and column deterioration that occur with Lewisite. The extracts were analysed by GC-AED and GC-MS with both electron and chemical ionization [63]. More GC-MS techniques will be described in Chapter 4 with respect to the numerous degradation products of Sulfur Mustard.

Although their volatility is fairly low, sternutator agents are easily analysed by gas chromatography. A variety of derivatizing agents and also detectors have been used for their analysis. Schoene and coworkers employed thioglycolic acid methyl ester as a derivatizing agent in order to speciate arsenic containing chemical warfare agents, including Clark I. They separated and monitored the analytes using GC coupled to a mass spectrometer and an atomic emission detector. Schoene *et al.* were able to successfully derivatize all analytes except for Adamsite. They were also able to obtain percentage yields ranging from 37 to 103% for the analytes tested [64].

1-Propanethiol has also been used as a derivatizing agent for arsenic-containing chemical warfare agents. Tørnes et al. collected sediment samples from shipwrecks that were known to hold CWAs. The sediment samples were then taken back to the laboratory, prepared and derivatized using various thiols and then analysed using GC-MS. 1-Propanethiol gave the highest overall yields compared with the other agents with yields ranging from 11 to 73%. The LOQ for the Clark I 1-propanethiol derivative was determined to be $6\,\mu g\,kg^{-1}$. Tørnes et al. were able to identify CWAs in the samples from the wreck sites, including the arsine oil mix that was used by Germany in WW II [65].

Clark I and Clark II have also been derivatized using other mercaptans and dimercaptans. Haas et al. were able to successfully derivatize Clark I and II using thioglycolic acid methyl ester, thioglycolic acid ethyl ester, 1-ethanthiol, 1-propanethiol, 1,2-ethanedithiol and 1,3-propanedithiol. These derivatives could also be identified using GC coupled to an electron capture device. The LODs for the derivatives ranged from 0.4 to 0.9 ng. Adamsite could not be detected by GC-ECD; however Hass et al. employed HPLC (high-performance liquid chromatography) as another separation technique for the CWAs. HPLC was used successfully for separating out Adamsite from a mixture, but the separation of Clark I and II was not possible. The authors concluded that GC-ECD gives a lower limit of detection and separation of Clark I and II, but Adamsite cannot be detected, so the selection of the chromatography depends greatly on the task at hand [66].

A combination of GC-AED and GC-MS has been used to successfully identify Adamsite. Schoene *et al.* took advantage of both elemental and molecular mass spectrometry to test two different derivatization methods for the analyte. The first derivatization method consisted of bromination of Adamsite by boiling in glacial acetic acid, forming 2,2′, 4, 4′, 6, 6′-hexabromodiphenylamine (HBDP). The second method was converting Adamsite into 10-ethyl-5,10-dihydrophenarsazine (EtPA) with N, N-dimethylformamide diethylacetal inside the hot GC injector. Through the use of GC-AED and GC-MS, Schoene and colleagues were able to determine derivatization yields of 90% for the bromination (20–80% in spiked soil samples) and 51% for the conversion method (53% in spiked soil samples). Owing to the sensitivities in sample and reaction variations of the bromination method, the conversion method was deemed to be optimal [67].

3.3.4 Liquid Chromatography (LC)/Liquid Chromatography-Mass Spectrometry (LC-MS)

Liquid chromatography can also be used as a separation technique for nerve agents. LC is a promising avenue worth pursuing because there is no need for derivatization steps, as required in GC, which can be difficult and time consuming, thus making the direct analysis of extracts with little or no sample preparation in aqueous matrices possible by LC. It is known that aqueous samples tend to hinder the derivatization process required for analysis by GC. Polar compounds, such as the alkylphosphonic acid degradation products of the nerve agents, are more suitable for LC allowing for the analysis of liquid-phase samples, which may not be detectable or are missed by GC. As with GC, LC is coupled with a selective detector (FPD) for identification and quantification [69].

Micro-LC, in combination with FPD, has also been pursued as a technique for the detection of CWAs [70, 71]. This technique proved to be very suitable for chiral separation of VX, which was not previously possible by GC [72].

Most notably, LC-MS is widely used for characterization and quantitation of nerve agents and hydrolysis products [60, 72–76]. After LC separation, several ionization techniques have been investigated for the detection of nerve agents, including early studies involving particle beam ionization (PBI), thermospray (TS) and continuous flow fast atom bombardment (FAB) [77–79]. Modern applications today have evolved from using TS and FAB techniques to the use of electrospray ionization (ESI), atmospheric pressure ionization (API) and atmospheric pressure chemical ionization (APCI).

LC-MS is a beneficial alternative to GC-MS, primarily when aqueous matrices are involved. As nerve agents undergo hydrolysis in environmental matrices, the detection of their degradation products is just as important as the detection of the nerve agents themselves, because degradation products strongly suggest the prior existence of CWA nerve agents. LC-MS can not only detect the parent nerve agents in aqueous environmental samples, but also the nonvolatile hydrolysis products in, potentially, a single analysis, nullifying the need for the further sample preparation and derivatization steps that are required in GC.

Determination of nerve agents and their degradation products in various environmental samples, such as aqueous, snow and soil, and even common office media samples, were performed using LC-ESI and μ-LC–ESI with MS, MS/MS or a time-of-flight MS [80–83]. Micro-LC procedures enhance the sensitivity needed for LC-MS instrumentation. Even with ESI-MS, LC-MS methods still have issues obtaining sensitivities comparable to those for GC-MS. MS/MS interfaced with LC-ESI does enhance sensitivity, providing increased signal-to-noise ratios and better selectivity for identification. Advances in API and APCI techniques, interfaced with quadrupole or ion trap MS systems, have also helped to overcome some of these disadvantages. API-MS is suitable for a broad range of compounds, from polar to nonpolar, as well as a wide range of flow rates and mobile phase compositions. Although field portable instrumentation is not yet fully developed, there is interest in LC-API-MS being used in a field-portable role similar to GC-MS [84].

3.3.5 Desorption Electrospray Ionization and Direct Analysis in Real Time Mass Spectrometry

An emerging technique in the field of chemical analysis is desorption electrospray ionization (DESI-MS/MS), which is a novel mass spectrometric technique that allows for direct sample ionization and analysis of surfaces, (Cooks and coworkers) [85]. DESI works by producing charged droplets in the solvent, which are electrosprayed onto the surface of interest, desorbing and ionizing the analyte of interest. Ionized molecules may then be detected by mass spectrometry. The detection and identification of nerve agents was possible using this technique [83, 86]. DESI is desirable because it would essentially result in quicker sample throughput. Although this is a new area of research, there is a promising future for a technique, such as DESI, to replace existing GC-MS and LC-MS methods for the analysis of nerve agents and degradants.

Similar to DESI, direct analysis in real time (DART) is another relatively new technique used to quickly and reliably analyse CWA nerve agents in the field quickly and reliably. DART is an atmospheric pressure technique that does not require vapor pressure, extractions, sprays, or sample preparation, and is nondestructive to the samples. It is accurate and fast for detecting and identifying (both qualitatively and quantitatively) chemical nerve agents in the field [87–90]. For all of these reasons, DART has some advantages however it can't analyzed mixtures which is a major drawback.

Recent developments and trends reviewed here reveal the amount of work and progress achieved in the field of analytical chemistry involved in the detection of CWAs. Analysis techniques are important and essential tools to be used in support of a defense against an attack using CWA nerve agents. As technology continues to improve, so will quantitative and qualitative limits, the time required for analysing samples, cost and the robustness and portability of instrumentation designed to target and detect analytes of interest.

REFERENCES

[1] Grate, J.W., Rose-Pehrsson, S.L., Venezky, D.L., et al. (1993) Anal. Chem., 65, 1868–1881.
[2] Grate, J.W., McGill, R.A. (1995) Anal. Chem., 67, 4015–4019.
[3] Shi, Y.B.,. Xiang, J.J., Feng, Q.H. et al. (2006) J. Phys.: Conf. Ser., 48, 292–297.
[4] Twomey, T.R., A. J. Caffrey, A.J., Chichester, D.L. (2007) INL Report, INL/CON-07-12304, February 2007.
[5] Bach, P., Ma, J.L., Froment, D., Jaureguy, J.C. (1993) Nucl. Instrum. Meth. Phys. Res. B: Beam Interact. Mater. Atoms, 79, 605–610.
[6] Anikiev, V.V., Kolesov, G.M. (2000) J. Anal. Chem., 55, 719–723.
[7] Arroyo, C.M., Sankovich, J.M., Burman, D.L., et al. (2005) Application of NMR and EPR Spectroscopy to the Analysis of the Reaction of Phosphovanadomolybdate Polyoxometalate ($H_5P^V_2Mo_{10}O_{40}$) with Chloroethyl Sulfides (Half-Sulfur Mustard and Sulfur Mustard), J. Med. CBR Def., 3, http://www.jmedcbr.org/Issue_0301/Arroyo/Arroyo_08_05.htm (accessed 12 December, 2010).
[8] Singh, V.V., Gupta, G., Sharma, R., et al. (2009) Synth. Met., 159, 1960–1967.

[9] Mesilaakso, M., Tolppa, E., Nousiainen, P. (1997) *Appl. Spectrosc.*, **51** (5), 733–737.

[10] Chemical and Biological Attacks, Detection & Response FAQ (2010), http://www.ki4u.com/chemical_biological_attack_detection.htm (accessed 12 December 2010).

[11] Nishiyama, K., Yamada, H., Matsuuraa, H. *et al.* (2008) 214th ECS Meeting, Honolulu, HI, October 2008.

[12] Berger, T., Ziegler, H., Krausa, M. (2000) Development of electrochemical sensors for the trace detection of explosives and for the detection of chemical warefare agents. SPIE Conference: Detection and Remediation Technologies for Mine and Mine-like Targets V, Orlando, FL, 2000, pp. 452–461.

[13] Davis, G.L. (2008) *CBRNE - Chemical Detection Equipment*, http://emedicine.medscape.com/article/833933-overview (accessed 12 December, 2010).

[14] Rostker, B. (1999) Information Paper, M256 Series Chemical Agent Detector Kit, http://www.gulflink.osd.mil/m256/ (accessed 12 December 2010).

[15] Hooijschuur, E.W.J., Kientz, C.E., Brinkman, U.A.T. (2002) *J. Chromatogr. A*, **982**, 177–200.

[16] Robins, W.H., Wright, B.W. (1994) *J. Chromatogr. A*, **680**, 667.

[17] Oehrle, S.A., Bossle, P.C. (1995) *J. Chromatogr. A*, **692**, 247.

[18] Cheicante, R.L., Stuff, J.R., Durst, H.D. (1995) *J. Chromatogr. A*, **711**, 347.

[19] Mercier, J.-P., Morro, P., Dreux, M., Tambute, A. (1996) *J. Chromatogr. A*, **741**, 279.

[20] Mercier, J.-P., Morin, P., Dreux, M., Tambute, A. (1997) *J. Chromatogr. A*, **779**, 245.

[21] Kientz, C.E., Hooijschuur, E.W.J., Brinkman, U.A.T. (1997) *J. Microcolumn Sep.* **9**, 253.

[22] Nassar, A.-E. F., Lucas, S.V., Jones, W.R., Hoffland, L.D. (1998) *Anal. Chem.*, **70**, 1085.

[23] Nassar, A.-E.F., Lucas, S.V., Myler, C.A. *et al.* (1998) *Anal. Chem.*, **70**, 3598.

[24] Rosso, T.E., Bossle, P.C. (1998) *J. Chromatogr. A*, **824**, 125.

[25] Mercier, J.-P., Chaimbault, P., Morina, P. *et al.* (1998) *J. Chromatogr. A*, **825**, 71.
[26] Nassar, A.-E.F., Lucas, S.V., Hoffland, L.D. (1999) *Anal. Chem.*, **71**, 1285.
[27] Zi-Hui, M., Qin, L. (2001) *Anal. Chim. Acta*, **435**, 121.
[28] Melanson, J.E., Boulet, C.A., Lucy, C.A. (2001) *Anal. Chem.*, **73**, 1809.
[29] Hooijschuur, E.W.J., Kientz, C.E., Brinkman, U.A.T. (2001) *J. Chromatogr. A*, **928**, 187.
[30] Melanson, J.E., Wong, B.L.-Y., Boulet, C.A., Lucy, C.A. (2001) *J. Chromatogr. A*, **920**, 359.
[31] Jiang, J., Lucy, C.A. (2002) *J. Chromatogr. A*, **966**, 239.
[32] Copper, C.L., Collins, G.E. (2004) *Electrophoresis*, **25**, 897.
[33] Smiths Detection Technical Information Guide CAM (Chemical Agent Monitor) (2010) http://www.smithsdetection.com/media/CAM_VT_EN_95593009.pdf (accessed 12 December 2010).
[34] Smiths Detection Technical Information Guide ECAM (Enhanced Chemical Agent Monitor) (2005) http://trace.smithsdetection.com/Documents/DataSheets/SDPL-5_Smiths_ECAM.pdf (accessed 12 December 2010).
[35] Smiths Detection Technical Information Guide CAM-2 (Chemical Agent Monitor) (2005) http://trace.smithsdetection.com/Documents/DataSheets/SDPL-1_Smiths_CAM-2.pdf (accessed 12 December 2010).
[36] Smiths Detection Technical Information APD 2000 (Advanced Portable Chemical Agent Detector) (2010) http://www.smithsdetection.com/APD2000.php (accessed 12 December 2010).
[37] Drager Multi-IMS – A Short Introduction Into The Operation, (2003) http://www.premium64.hu/doksik/multiims_en.pdf (accessed 12 December 2010.)
[38] Smiths Detection Technical Information Sabre 4000 (2010) http://www.smithsdetection.com/SABRE_4000.php (accessed 12 December 2010).
[39] Longworth, T.L., Ong, K.Y. (2001) Domestic Preparedness Program: Testing of Sabre 2000 Handheld Trace and Vapor Detector Against Chemical Warfare Agents Summary Report, Aberdeen Proving Ground, Maryland, Report number: 0704–0188, pp 1–28.

[40] Davis, G.L. (2008) CBRNE - Chemical Detection Equipment, http://emedicine.medscape.com/article/833933-overview (accessed 12 December 2010).

[41] Modi A, Koratkar, N., Lass E. *et al.* (2003) *Nature*, **424**, 171–174.

[42] Brletich, N.R., Waters, M.J., Bowen, G.W., *et al.* (1995) *Worldwide Chemical Detection Equipment Handbook*, US Department of Defense.

[43] Department of Homeland Security (2007) *Guide for the Selection of Chemical Detection Equipment for Emergency First Responders*, 3rd edn, January 2007, Department of Homeland Security, Coordinated by NIST-OLES.

[44] Fitch, J.P., Raber, E., Imbro, D.R. (2003) *Science*, **302**, 1350–1354.

[45] Institute of Medicine (1999) *Chemical and Biological Terrorism: Research and Development to Improve Civilian Medical Response*, National Academies Press, pp. 37–52.

[46] Zajtchuk, R.M.C., Bellamy, R.F. (1997) Chemical defense equipment, in *Textbook of Military Medicine: Medical Aspects of Chemical and Biological Warfare*, Office of the Surgeon General.

[47] Zimmermann, S., Barth, S., Baether, W.K.M., Ringer, J. (2008) *Anal. Chem.*, **80**, 6671–6676.

[48] Kagakkai, N.B. (2010) *Bunseki Kagaku*, **59**, 65–76.

[49] Kientz, C.E. (1998) *J. Chromatogr. A*, **814**, 1–23.

[50] Zhao, C., Chen, Z., Xu, S., Wei, C. (2008) *Fenxi Ceshi Xuebao*, **27**, 1237–1240.

[51] Tang, H., Cheng, Z., Zhu, H. *et al.* (2008) *Appl. Catal. B*, **79**, 323–333.

[52] Li, W., Zuo, B., Zhang, T., Li, S. (2006) *Huanjing Wuran Yu Fangzhi*, **28**, 796–798.

[53] Mazurek, M., Witkiewicz, Z., Popiel, S., Sliwakowski, M. (2001) *J. Chromatogr. A*, **919**, 133–145.

[54] Lattin, F.G., Mehta, U.J., Jakubowski, E.M. *et al.* (1996) *An analytical method for determination of low level airborne exposure levels to lewisite/mustard*, National Technical Information Service, pp. 177–180.

[55] Shearer, R.L., O'Neal, D.L., Rios, R., Baker, M.D. (1990) *J. Chromatogr. Sci.*, **28**, 24–28.

[56] Juillet, Y., Gibert, E., Begos, A., Bellier, B. (2005) *Anal. Bioanal. Chem.*, **383**, 848–856.

[57] Albro, T.G., Lippert, J. (1996) *Use of the atomic emission detector for screening and detection of chemical warfare agents and their breakdown products*, National Technical Information Service, pp. 171–176.

[58] Black, R.M. (2010) *J. Chromatogr. B: Anal. Technol. Biomed. Life Sci.*, **878**, 1207–1215.

[59] Black, R.M., Muir, B. (2003) *J. Chromatogr. A*, **1000**, 253–281.

[60] Hooijschuur, E.W.J., Kientz, C.E., Brinkman, U.A.T. (2002) *J. Chromatogr. A*, **982**, 177–200.

[61] Stuart, D.A., Biggs, K.B., Van, D.R.P. (2006) *Analyst (Cambridge, UK)*, **131**, 568–572.

[62] Hu, X., Zhou, Y., Feng, C. *et al.* (1999) *Junshi Yixue Kexueyuan Yuankan*, **23**, 215–217.

[63] Down, S. (2005) *GC/MS Targets Abandoned Chemical Weapons Shells*, http://www.separationsnow.com/coi/cda/detail.cda;jsession id=64166836840AB0DD62988D53D3A6041F?id=11452& type=Feature&chId=3&page=1 (accessed 21 September 2010).

[64] Schoene, K., Steinhanses, J., Bruckert, H.J., König, A. (1992) *J. Chromatogr. A*, **605**, 257–262.

[65] Tørnes, J.A., Opstad, A.M., Johnsen, B.A. (2006) *Sci. Total Environ.*, **356**, 235–246.

[66] Haas, R., Schmidt, T.C., Steinbach, K., von Löw, E. (1998) *Fresenius J. Anal. Chem.*, **361**, 313–318.

[67] Schoene, K., Bruckert, H.J., Jürling, H., Steinhanses, J. (1996) *J. Chromatogr. A*, **719**, 401–409.

[68] Brickhouse, M.D., Creasy, W.R., Williams, B.R., *et al.* (2000) *J. Chromatogr. A*, **883**, 185.

[69] Hooijschuur, E.W.J., Kientz, C.E., Brinkman., U.A.Th. (2001) *J. Chromatogr. A*, **907**, 165.

[70] Kientz, C.E., Verweij, A., de Jong, G.J., Brinkman, U.A.Th. (1992) *J. Microcol. Sep.*, **4**, 465.

[71] Kientz, C.E., Verweij, A., de Jong, G.J., Brinkman, U.A.Th. (1992) *J. Microcol. Sep.*, **4**, 477.

[72] Kientz, C.E., Langenberg, J.P., Brinkman, U.A.Th. (1994) *J. High Resolut. Chromatogr.*, **17**, 95.

[73] D'Agostino, P.A., Hancock, J.R., Provost, L.R. (2001) *Adv. Mass Spectrom.*, **15**, 297.

[74] Black, R.M. and Read, R.W. (2000) Liquid chromatography/mass spectrometry in analysis of chemicals related to the chemicals weapons convention, in *Encyclopedia of Analytical Chemistry*, (ed. R.A. Meyers), John Wiley & Sons, Ltd, Chichester, pp. 1007–1025.

[75] Noort, D., Benschop, H.P., Black, R.M., *Toxicol. App. Pharmacol.*, **184** (2002) 116.

[76] Ellis-Steinborner, S., Ramachandran, A., Blanksby, S.J. (2006) *Rapid Commun. Mass Spectrom.*, **20**, 1939.

[77] Niessen, W.M.A., Van der Greef, J. (1992) *Liquid Chromatography–Mass Spectrometry*, Marcel Dekker, New York, 1992.

[78] Wils, E.R.J., Hulst, A.G. (1990) *J. Chromatogr.*, **523**, 151.

[79] Wils, E.R.J., Hulst, A.G. (1992) *Fresenius J. Anal. Chem.*, **342**, 749.

[80] D'Agostino, P.A., Hancock, J.R., Provost, L.R. (1999) *J. Chromatogr. A*, **840**, 289.

[81] D'Agostino, P.A., Hancock, J.R., Provost, L.R. (2001) *J. Chromatogr. A*, **912**, 291.

[82] D'Agostino, P.A., Chenier, C.L., Hancock, J.R. (2002) *J. Chromatogr. A*, **950**, 149.

[83] D'Agostino, P.A., Hancock, J.R., Chenier, C.L., Lepage, J. (2006) *J. Chromatogr. A*, **1110**, 86.

[84] Smith, J.R., Shih, M.L., Price, E.O. *et al.* (2001) *J. Appl. Toxicol.*, **21**, S35.

[85] Takats, Z., Wiseman, J.M., Gologan, B., Cooks, R.G. (2004) *Science*, **306**, 471.

[86] D'Agostino, P.A., Chenier, C.L., Hancock, J.R., Lepage, (2007). *Rapid Commun. Mass Spectrom.*, **21**, 543.

[87] Nilles, J.M., Connell, T.R., Durst, H.D. (2009) *Anal. Chem.*, **81**, 6744.

[88] Laramé, J.A., Cody, R.B., Nilles, J.M., Durst, H.D. (2007) Forensic applications of DART (direct analysis in real time) mass

spectrometry, in *Forensic Analysis on the Cutting Edge*, (ed. R.D. Blackledge) John Wiley & Sons, Inc., Hoboken, pp. 175–195.

[89] Cody, R.B., Laramé, J.A., Durst, H.D. (2005) *Anal. Chem.*, **77**, 2297.

[90] Laramé, J.A., Durst, H.D., Connell, T.R., Nilles, J.M. (2008) *Am. Lab.*, **40**, 16.

4

Chemical Warfare Agent Degradation Products

Analysis of Chemical Warfare Degradation Products, First Edition. Karolin K. Kroening,
Renee N. Easter, Douglas D. Richardson, Stuart A. Willison and Joseph A. Caruso.
© 2011 John Wiley & Sons, Ltd. Published 2011 by John Wiley & Sons, Ltd.

4.1 ANALYSIS OF NERVE AGENT DEGRADATION PRODUCTS

The Organization for the Prohibition of Chemical Weapons (OPCW) was established by member states following the Chemical Weapons Convention (CWC) of 1997 [1]. The mission of the OPCW is to promote and enforce an end to the production, acquisition and direct or indirect transfer of chemical weapons, in addition to mandating the destruction of all chemical weapons held in reserve for member states. Membership as of September 2009 in the OPCW consisted of 188 member states, two signatory states and five non-signatory states, corresponding to 98% of the global population [1]. Since the inception of OPCW (April 1997) more than 3900 inspections have taken place at 195 chemical weapon-related sites and 1103 industrial sites on the territory of 81 member states [1]. To date, approximately 55% of the declared chemical agent stockpiles have been destroyed worldwide, with more than 5000 industrial facilities liable to inspection [1].

These toxic chemicals, which include Schedule 1 blood, choking, blistering and nerve agents, threaten not only the global population but also important food and environmental resources. As a consequence of the more than 30 000 metric tonnes of declared worldwide reserves and due to the threat of terrorist attacks, development of robust analytical techniques for the analysis of chemical warfare agents (CWA) directly, or through the trail left by their degradation metabolites, is of vital importance to bolster our capabilities of dealing with CWA storage/destruction.

Formation of alkyl phosphonic acid degradation products in the environment follows a highly specific hydrolysis pathway from the parent dialkyl phosphonate ester CWA's. Nerve agent hydrolysis products are unique compared with other organophosphorus hydrolysates, such as pesticides and herbicides, due to the carbon–phosphorus bond intrinsic to the parent CWAs [2]. This unique carbon–phosphorus chemistry provides selective, less toxic chemical species as an alternative reference material for the detection of CWAs. Common alkyl phosphonic acid degradation products include: ethyl methylphosphonic acid (EMPA, VX), isobutyl hydrogen methylphosphonate (IBHMP, RVX), isopropyl methylphosphonic acid (IMPA, Sarin), cyclohexyl methylphosphonic acid (CMPA, Cyclosarin), ethyl hydrogen dimethylamidophosphate sodium salt (EHDAP, Tabun), pinacolyl methylphosphonic acid (PMPA, Soman) and methylphosphonic acid (MPA). A summary of alkyl phosphonic acid degradation products and their corresponding parent CWA can be found in Figure 4.1.

Alkyl phosphonic acid nerve agent degradation products present difficulties for sample preparation and ultra-trace analysis due to their high polarity, low volatility and lack of a good chromophore for ultraviolet (UV) detection.

4.1.1 Sample Preparation

Common sample matrices associated with exposure to and release of nerve agents and their corresponding alkyl phosphonic acid degradation products are typically limited to armaments (military), environmental (i.e., soil, water, food), synthetic (clothing, plastics), or biologic (blood, saliva, urine). Owing to the complex nature of these sample matrices, no universal sample preparation technique exists for alkyl phosphonic acid degradation products. A variety of sample preparation technologies have been developed to obtain samples compatible with commonly used analytical techniques. Nevertheless, detailed knowledge of the sample matrix and analysis technique typically determines the

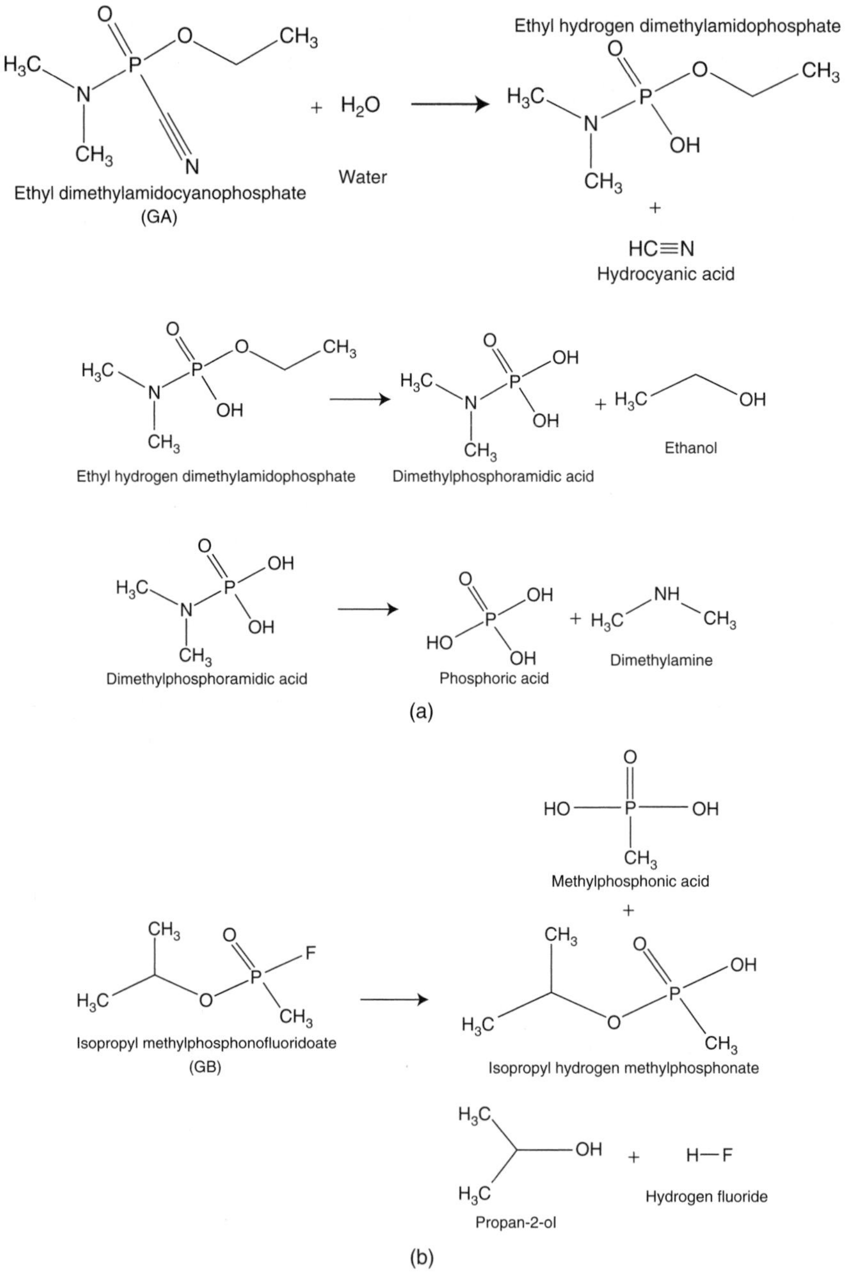

(a)

(b)

Figure 4.1 Hydrolysis pathways for: (a) GA (Tabun); (b) GB (Sarin); (c) GF (cyclosarin); (d) GD (Soman); and (e) VX

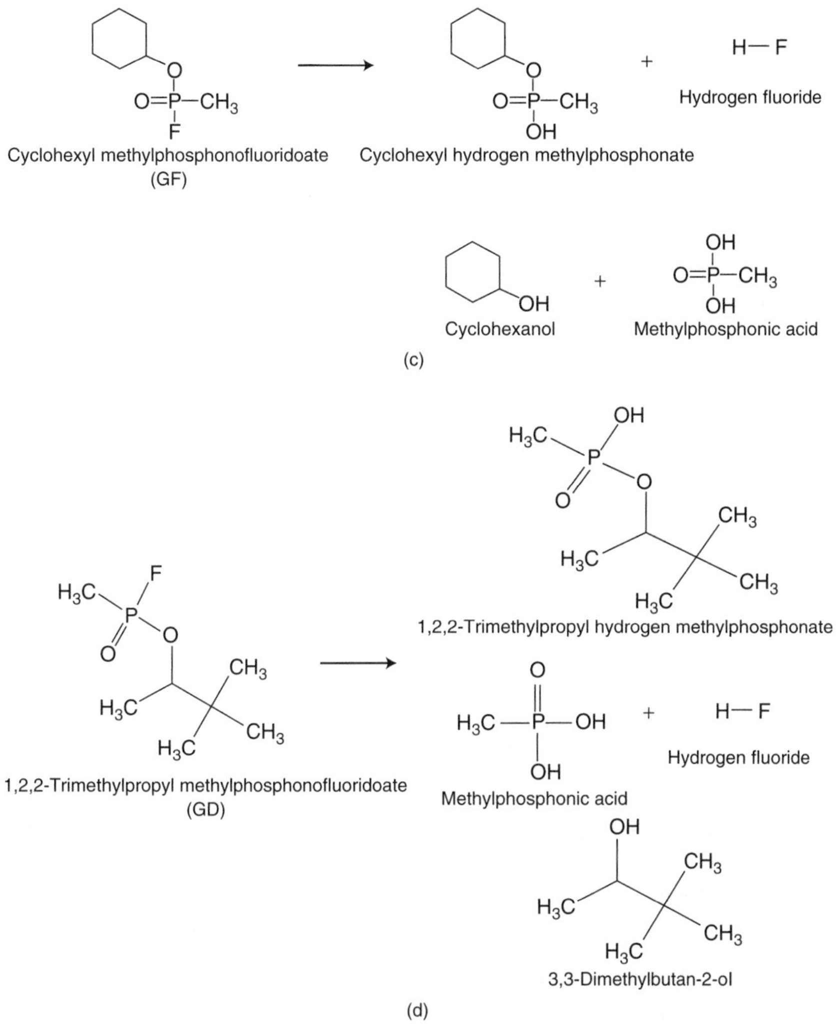

(c)

(d)

Figure 4.1   (*Continued*)

best procedure or combination for sample preparation. Typical sample preparation methods include: liquid–liquid extraction, solid phase extraction (SPE), solid phase microextraction (SPME), stir bar sorptive extraction (SBSE) and derivatization.

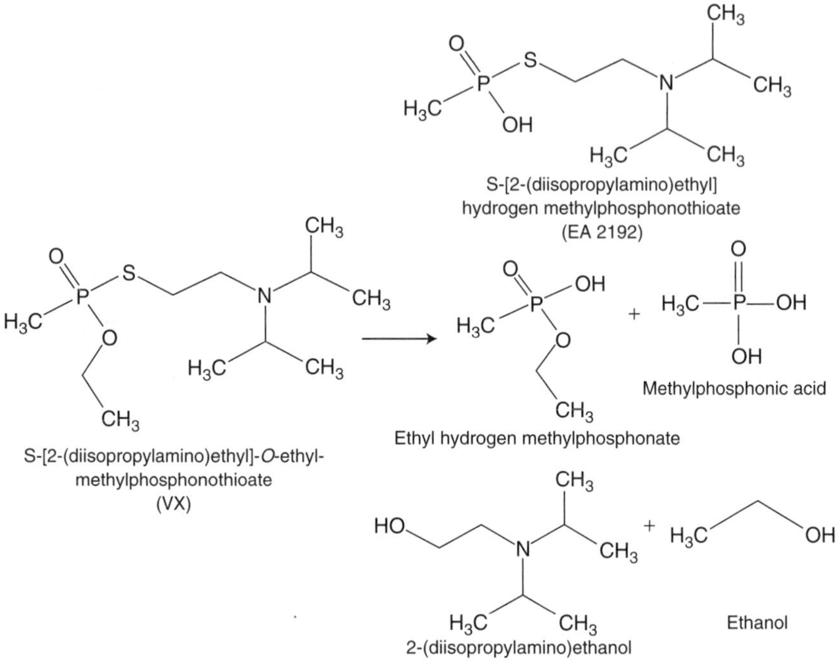

S-[2-(diisopropylamino)ethyl]
hydrogen methylphosphonothioate
(EA 2192)

S-[2-(diisopropylamino)ethyl]-*O*-ethyl-
methylphosphonothioate
(VX)

Ethyl hydrogen methylphosphonate

Methylphosphonic acid

2-(diisopropylamino)ethanol

Ethanol

(e)

Figure 4.1 (*Continued*)

4.1.2 Liquid–Liquid Extraction (Pre-concentration)

Liquid–liquid extraction (LLE) is one of the oldest and most common types of sample preparation utilized in analytical chemistry. This extraction technique is based upon the solubility/affinity of an analyte between two immiscible liquids, typically composed of one aqueous and one organic layer. In a recent review, John *et al.* [3] discussed the application of LLE for the sample preparation of alkyl phosphonic acids in a variety of sample matrices including water, serum and acidified urine. Modifications to typical LLEs have resulted in miniaturization, preconcentration and added selectivity through introduction of ion-pairing agents. Xu *et al.* [4] developed an ion-pairing liquid–liquid–liquid microextraction (LLLME) technique for the analysis of a variety of alkyl phosphonic acid degradation products. LLLME consists of an initial aqueous extraction, then extraction into an organic solvent, followed again by a back-extraction into the aqueous phase for analysis. Owing to the small volumes associated with the aqueous fractions of the LLLME technique, analyte preconcentration is achieved prior to analysis.

4.1.3 Solid Phase Extraction (SPE)

Similar to liquid chromatography, solid phase extraction (SPE) is the sample preparation technique of separating analytes in solution or suspension (mobiles phase) by passing over a solid phase (stationary) of specific chemistry. Analyte preconcentration/purification is achieved through retention or elution away from the interfering matrix components. Typical examples of sorbents utilized for solid phase extraction applications include: ion exchange (cation and anion), normal phase (hydrophilic), reversed phase (hydrophobic), polymeric (combination), or molecularly imprinted polymer. An example of SPE for alkyl phosphonic acid degradation products from Le Moullec et al. [5] utilized molecularly imprinted polymers for the selective preconcentration of EMPA from soil extracts with 93% recovery. Alternatively, ion-pairing SPE has been utilized for the methyl, ethyl and propyl phosphonic acid degradation products with greater than 90% recovery [6]. Other work has demonstrated the use of strong anion-exchange (Oasis Max) cartridges for the successful preconcentration of alkyl phosphonic acid degradation products in soil samples from OPCW proficiency test samples [7]. More recent works involving OPCW proficiency testing have demonstrated the stability of multiple alkyl phosphonic acid degradation products on strong anion exchange disks over a 36 day period [8].

4.1.4 Solid Phase Microextraction (SPME)

Solid phase microextraction (SPME) is a simple, portable, inexpensive and solvent free preconcentration method, which allows sorption of analytes onto the surface of a modified fused microfiber from a variety of sample matrices. Optimization of the analyte sorption process with SPME is based upon equilibrium between the analyte in the sample (solution or headspace) and on the fiber, requiring consideration of the fiber stationary phase, pH, ionic strength, sampling time, agitation and desorption temperature. These parameters can vary widely depending upon analyte polarity, volatility and size, resulting in a competitive sorption process. Reardan and Harrington [9] successfully demonstrated the use of SPME fibers for the analysis of alkyl phosphonic acid degradation products from soil. Additionally, Zygmunt *et al.* [10] reviewed the use of SPME for sample preparation of alkyl phosphonic acid degradation products.

4.1.5 Stir Bar Sorptive Extraction (SBSE)

Following on from the same equilibrium principles of SPME, stir bar sorptive extraction provides an alternative sample preparation technique for alkyl phosphonic acid degradation products [3]. The stir bar utilized for sampling contains a thicker coating and larger surface area compared with SPME fibers, resulting in longer equilibration times and a greater degree of preconcentration. Lancas *et al.* [11] recently reviewed the applications of SBSE for sample preparation and preconcentration techniques for modern analytical analysis.

4.1.6 Derivatization

Owing to the nonvolatile nature of these moderate to highly polar alkyl phosphonic acid degradation products, multiple types of derivatization techniques, including pentafluorobenzyl esters, methyl esters, trimethylsilyl and *tert*-butyldimethylsilyl esters, have been studied for the generation of more volatile species amenable to GC separation. Black and Muir [12] reviewed derivatization reactions for the analysis of chemical warfare agents and their degradation products. This review described both trimethylsilyl (TMS) and *tert*-butyldimethylsilyl (TBDMS) derivatives, the most popular for alkyl phosphonic acids, with the TBDMS derivatives suspected of being the most stable and least sensitive to moisture [12]. Figure 4.2 describes the TBDMS

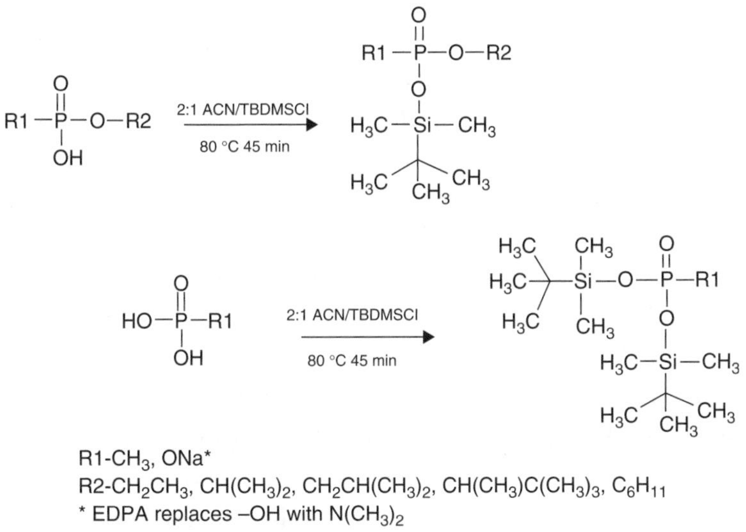
2:1 ACN/TBDMSCl
80 °C 45 min
2:1 ACN/TBDMSCl
80 °C 45 min
R1-CH3, ONa*
R2-CH2CH3, CH(CH3)2, CH2CH(CH3)2, CH(CH3)C(CH3)3, C6H11
* EDPA replaces –OH with N(CH3)2

Figure 4.2 TBDMS esterification of alkyl phosphonic acids [14]. Reprinted from Richardson, D.D., Caruso, J.A. Caruso, *Analytical and Bioanalytical Chemistry* 2007, 389, 679–682, with kind permission from Springer Science + Business Media

esterification of alkyl phosphonic acids. Post-column (liquid chromatography) derivatization has been utilized through formation of the methyl ester derivatives of alkyl phosphonic acid degradation products of CWA [13]. Richardson and Caruso [14] have described the use of TBDMS derivatives of alkyl phosphonic acid degradation product extracts from soil, river water and in the presence of common organophosphorus pesticide mixtures. Recently, Subramaniam *et al.* [8] utilized in-vial solid phase derivatization for the formation of the TMS derivatives of alkyl phosphonic acid degradation products from aqueous OPCW proficiency test samples.

4.2 ANALYTICAL TECHNIQUES

A variety of analytical techniques have been utilized for the analysis of alkyl phosphonic acid degradation products of nerve agents. Physical and chemical properties associated with alkyl phosphonic acid degradation products has allowed for the use of routine separation techniques, such as liquid chromatography (LC), gas chromatography (GC), capillary electrophoresis (CE) and ion mobility (IM), in conjunction with a variety of detectors [3, 15]. Mass spectrometry (MS) or tandem mass spectrometry (MS/MS) are the most commonly used detection technique coupled with analytical separation for the analysis of alkyl phosphonic acid degradation products. Other suitable detectors include: nuclear magnetic resonance (NMR), Fourier transform infra-red (FTIR) spectroscopy, flame ionization detector (FID), flame photometric detector (FPD) and a nitrogen–phosphorus detector (NPD) [3, 15, 16]. An overview of the most routine analytical techniques for the analysis of alkyl phosphonic acid degradation products of nerve agents is provided below.

4.2.1 Gas Chromatography (GC)

Gas chromatography coupled with mass spectrometric detection is the most commonly used detection technique for the analysis of parent organophosphorus nerve agents, because of their intrinsic volatility. Conversely, alkyl phosphonic acid degradation products are nonvolatile, highly polar species, which require derivatization prior to analysis by GC-MS. GC-MS with electron impact ionization (EI) of derivatized alkyl phosphonic degradation products results in intrinsic fragmentation patterns for each species, which are useful in structural elucidation and for mass spectral library searches. Complementary to EI, chemical ionization (CI) with methane, isobutane or ammonia reagent gas provides intact molecular mass identification for GC-MS analysis.

Saradhi *et al.* [6] utilized GC-MS in scanning and single ion monitoring (SIM) modes with EI ionization for the analysis of multiple alkyl phosphonic acid degradation products in aqueous samples. Detections limits for this technique ranged from $0.08-0.5$ mg L^{-1} for the scanning mode to $5-50$ µg L^{-1} for SIM [6]. The resulting method was applied to aqueous OPCW proficiency test samples with recoveries ranging from 20 to 90% for alkyl phosphonic acid degradation products [6].

Kanaujia *et al.* [7] investigated the use of GC-MS for the analysis of soil samples from OPCW proficiency testing for alkyl phosphonic acid degradation products. Detection limits for these experiments ranged from $0.05-0.1$ µg mL^{-1} for full scan and $0.0005-0.005$ µg mL^{-1} for SIM, with recoveries of 90–95% for monobasic and 60–75% for dibasic alkyl phosphonic degradation products [7].

More recently, Subramaniam *et al.* [8] applied the use of in-vial solid phase derivatization with GC-MS for the analysis of multiple alkyl phosphonic acid degradation products from aqueous OPCW proficiency testing samples. Analytical figures of merit for these experiments included a detection limit of 0.14 ppb in the SIM mode with recoveries ranging from 83 to 101% [8]. Gas chromatography coupled with Fourier transform infra-red spectroscopy (GC-FTIR) has been described in the literature for the analysis of 55 nerve agent homologs, including alkyl phosphonic acid degradation products [17]. Further details on alkyl phosphonic acid degradation product analysis by GC-MS and related techniques can be found in the reviews by Hooijschuur *et al.* [16] and Zygmunt *et al.* [10].

4.2.2 Liquid Chromatography (LC)

Liquid chromatography coupled with mass spectrometric (LC-MS or MS/MS) detection has become one of the most widely used analytical techniques for the analysis of alkyl phosphonic acid degradation products. Figure 4.3 describes trends in sample preparation and ionization techniques for analysis of alkyl phosphonic acid degradation products by LC-MS.

The nonvolatile, highly polar chemistry of alkyl phosphonic acid degradation products provides ideal candidates for separation by routine liquid chromatographic techniques. These liquid separations are typically accomplished through the use of a reversed phase method where the mobile phase is polar (hydrophilic) and the stationary phase nonpolar (hydrophobic). Modified approaches to reversed phase LC include the use of ion and ion-pairing chromatography, providing excellent separation of charged species.

Alkyl phosphonic acid degradation products provide good ionization and fragmentation with MS and MS/MS detection. Typical ionization techniques coupled with LC are electrospray ionization (ESI) and atmospheric pressure

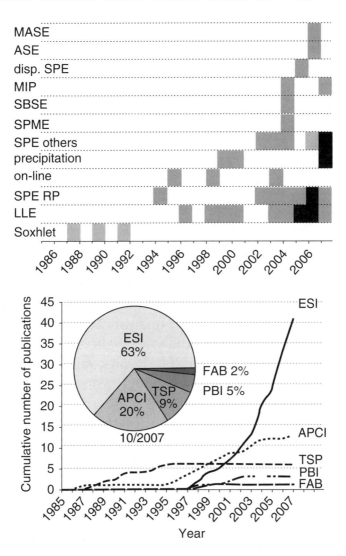

Figure 4.3 (*L*) Trends in sample preparation for analysis of alkyl phosphonic acid degradation products by LC-MS. (*R*) Trends in ionization techniques for analysis of alkyl phosphonic acid degradation products by LC-MS [3]. Reprinted from John, H., Worek, F., Thiermann, H., LC-MS-based procedures for monitoring of toxic organophosphorus compounds and verification of pesticide and nerve agent poisoning. *Analytical and Bioanalytical Chemistry* 2008, 391, (1), 97–116 with permission from Elsevier

chemical ionization (APCI). These ionization techniques are considered to be soft ionization, resulting in simple spectra of $[M+H]^+$ fragments. MS/MS provides an instrument based approach to induce fragmentation on parent MS ions, producing more sensitive daughter-ion spectra. Previously, Creasy [13] described the use of post-column derivatization within the APCI source with LC-MS for the analysis of methyl phosphonic acid (MPA). Drawbacks to this approach were attributed to contamination of the ionization source, requiring daily maintenance following analysis [13].

Isotope dilution tandem mass spectrometry with liquid chromatography (LC-MS/MS) was applied to the analysis of alkyl phoshphonic acid degradation products in urine [18]. Deuteriated reference standards of typical alkyl phosphonic acid degradation products provided unique internal standard MS/MS spectra allowing direct quantification of the selected degradation products [18]. Analytical figures of merit showed detection limits of less than 200 fg on-column for all alkyl phosphonic acid degradation products tested [18].

Mawhinney et al. [19] described an alternative hydrophilic interaction LC-MS/MS approach for the analysis of degradation products in urine. Detection limits for this approach ranged from 0.8 to 6 pg on-column for a variety of alkyl phosphonic acid degradation products [19]. Table 4.1 describes the error in the response ratios for various standard and quality control samples.

Additional work from the same group utilized the post-column addition of organic solvents for enhanced sensitivity in LC-MS/MS analysis of degradation products [20]. Sensitivity improvement factors of up to 60 were achieved with the post-column addition approach [20]. More recently, Owens and Koester [21] described the use of LC-MS/MS for the analysis of alkyl phosphonic acid degradation products in beverages. Method detection limits

Table 4.1 Percentage error for response ratios of alkyl phosphonic acid standard and quality control samples [19]. Reprinted from Mawhinney, D. B.; Hamelin, E. I.; Fraser, R.; Silva, S. S.; Pavlopoulos, A. J.; Kobelski, R. J., The determination of organophosphonate nerve agent metabolites in human urine by hydrophilic interaction liquid chromatography tandem mass spectrometry: Journal of Chromatography, B: Analytical Technologies in the Biomedical and Life Sciences 2007, 852, (1–2), 235–243 with permission from Elsevier

Concentration (ng mL^{-1})	Average error^a (%) ($N = 5$)				
	EMPA	IMPA	MMPA	PMPA	CMPA
1	6.3	4.4	5.3	4	2.7
2	6.3	1.9	5.9	8	5
5	5.6	8.7	2.9	2.1	3.2
10	3.6	2.8	2.6	2.2	1.5
15	3.9	4	1.7	2.7	3.9
25	2.4	4	2	2.5	2.8
50	2.8	2.9	1.3	3	2
75	2.2	2.5	1.7	1.9	2
100	3	1.9	1.4	2.6	1.7
20	1.5	3.6	1.7	2	2.1

^{a}EMPA is the metabolite of GB, MMPA is the metabolite of rVX, PMPA is the metabolite of GD and CMPA is the metabolite of GF.

of >0.02 ng on-column for a variety of alkyl phosphonic acid degradation products in water, apple juice, juice drink, cola and whole milk were demonstrated [21]. As an alternative to LC-MS, Mazumder *et al.* [22] developed a liquid chromatography–ultraviolet–nuclear magnetic resonance (LC-UV-NMR) approach to the analysis of alkyl phosphonic acid degradation products. Further information on LC and LC-MS methods for analysis of alkyl phosphonic acid degradation products can be found in the comprehensive reviews by Hooijschuu *et al.* [16] and John *et al.* [3].

4.2.3 Elemental Speciation

Over the past few decades a shift in the research focus for the analysis of metal- and nonmetal-containing samples has resulted in the evolution of elemental speciation. Elemental speciation is the development of species-specific analytical techniques for the differentiation between the essentiality and toxicity of these elemental species in environmental and biological systems. Elemental speciation analyses are performed by coupling analytical separation techniques, such as LC, GC and CE, to element specific detection with inductively coupled plasma mass spectrometry (ICP-MS). Element specific mass spectrometry with ICP-MS has become the benchmark technique for speciation analysis due to the sensitivity, selectivity, multielement detection capability, wide dynamic range and ease of chromatographic hyphenation that accompanies this detection system. Richardson et al. [23] utilized ion-pairing HPLC-ICP-MS for the analysis of alkyl phosphonic acid degradation products in river water and soil matrices. Analytical figures of merit included detection limits of $140-260 \, pg \, mL^{-1}$ and recoveries of 69–86% [23]. This work was the first use of elemental speciation with ^{31}P detection for the analysis of alkyl phosphonic acid degradation products [23]. Kubachka et al. [24] were the first to demonstrate the use of HPLC-ICP-MS for the analysis of alkyl phosphonic acid degradation products in beverages and lettuce extracts (Figure 4.4).

Moreover, GC-ICP-MS has also been utilized for the analysis of TBDMS derivatives of alkyl phosphonic acid degradation products in river water and soil extracts [25]. In addition, the same group investigated the analysis of TBDMS derivatives of alkyl phosphonic acid degradation products in the presence of common herbicides and pesticides [14]. Kroening et al. [26] recently reviewed the use

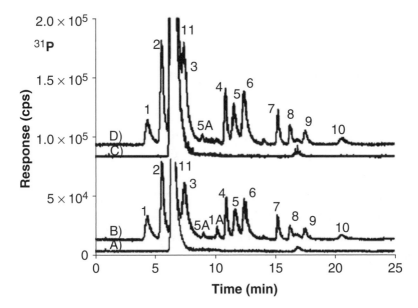

Figure 4.4 LC-ICP-MS analysis of alkyl phosphonic acids in food and beverage matrices [24]. Reprinted from Kubachka, K.M., Richardson, D.D., Heitkemper, D.T., Caruso, J.A., Detection of chemical warfare agent degradation products in foods using liquid chromatography coupled to inductively coupled plasma mass spectrometry and electrospray ionization mass spectrometer. *Journal of Chromatography A*, 2008, 1202, (2), 124–131, with permission from Elsevier

of elemental speciation for the analysis of alkyl phosphonic degradation products.

4.2.4 Ion Mobility

Ion mobility spectroscopy is based on the migration of a charged chemical species through a weak electric field in a gas phase. The mobility is calculated from the ion velocity within the drift tube of the ion mobility spectrometer. The rapid analysis time and portability of IMS make it a popular analytical technique for alkyl phosphonic acid degradation product analysis. Kanu *et al.* [27] demonstrated the use of thermal desorption IMS for the analysis of chemical warfare agent simulants and degradation products. TD-IMS is typically used for explosive or drug detection, but this work demonstrated excellent sensitivity of $3.2 \times 10^{-2}\,A\,g^{-1}$ with detection limits near 15 pg for common degradation products [27]. Figure 4.5 provides a cross-section diagram of a TD-IMS instrument.

In similar work, Rearden and Harrington [9] demonstrated the use of SPME-IMS for the analysis of alkyl phosphonic acid degradation products in soil samples. Detection limits of

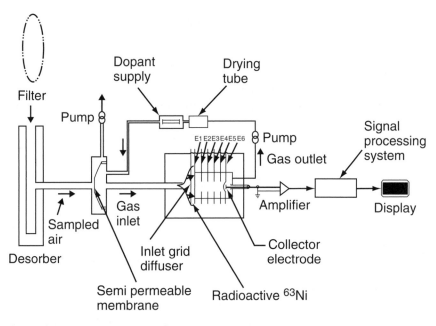

Figure 4.5 Cross-section diagram of a TD-IMS instrument [27]. Reprinted from Kanu, A.B., Haigh, P.E., Hill, H.H., Surface detection of chemical warfare agent simulants and degradation products. *Analytica Chimica Acta* 2005, 553, (1-2), 148–159, with permission from Elsevier

$10\,\mu g\,g^{-1}$ were achieved with an analysis time of 30 min [9]. More recently, Kolakowski *et al.* [28] demonstrated the use of atmospheric pressure ionization–high field asymmetric waveform ion mobility mass spectrometry for the analysis of nerve agent degradation products in food samples.

4.2.5 Capillary Electrophoresis

Capillary electrophoresis is one of the most popular analytical separation techniques for charged small molecules. Separation is achieved based upon the size-to-charge ratio of each analyte and their migration (electrophoretic mobility) through a modified silica capillary with high electric current. CE is the only analytical separation technique that can resolve positive, negative and neutral species in a single run. Lagarrigue *et al.* [29] demonstrated the use of CE–ion trap mass spectrometry for the analysis of isomeric acidic degradation products of organophosphorus chemical warfare agents. This work successfully demonstrated the detection and identification of alkyl phosphonic degradation products in soil extracts spikes of $5\,\mu g\,mL^{-1}$ [29]. Recently, Xu and coworkers [4, 30] implemented electrochemical detection by using CE with contactless conductivity detection for the analysis of nerve agent degradation products. Conductivity detection provided detection limits of $0.5–2.7\,ng\,mL^{-1}$ in river water and $0.09–0.44\,\mu g\,g^{-1}$ in soil for multiple alkly phosphonic acid degradation products [30].

4.3 ANALYSIS OF SULFUR MUSTARD DEGRADATION PRODUCTS

Chemical warfare agents have been produced in enormous amounts over the past 100 years and their disposal is very expensive. A safe disposal procedure requires secure packaging, appropriate handling with safety precautions carried out by experts, careful transportation of hazard material with many requisite documents [2]. As to be expected, the cost of their disposal is escalating.

Often CWAs are simply lying in the sea, where they were dumped years ago during the World Wars and their recovery has been unnecessary. However, worldwide, incidents of contamination from CWAs pollution are known. One example was in Bari, Italy, where fishermen were exposed to shells containing Sulfur Mustard from a bombardment on Bari's harbor during World War II. These shells had been lying in the sea for over 40 years and were still able to produce burning and almost fatal consequences [4].

These, and many more health concerns related to CWAs in the environment, are a good reason to develop methods that allow detection of CWAs at the lowest levels possible.

It is not just for these reasons that it is important that the parent agents mentioned in the previous chapters be identified, but so should their degradation products, because they too can have toxicities that are of concern [6]. Most of these parent agents undergo hydrolysis, giving rise to CWADPs (chemical warfare agent degradation products) whose toxicity can be alarming [7]. However, even though their toxicity is definitely lower than the original CWA, this may not always be the case and concentrations, exposure times, cell uptake effectiveness, and so on, will have a major influence, therefore their detection at the lowest level possible is also important.

When a small sample of soil or water is taken for analysis, it is extremely important to identify low levels (in the parts per trillion range) of these compounds and a fast detection technique is desirable.

An important hydrolysis degradation product of Sulfur Mustard is thiodiglycol (TDG), bis(2-hydroxyethyl)sulfide.

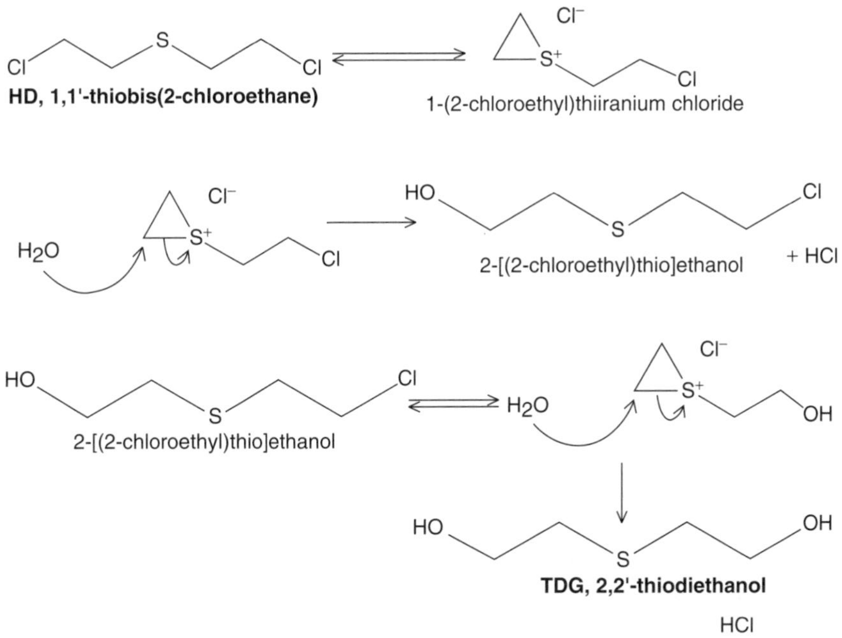

Figure 4.6 Main degradation hydrolysis products of Sulfur Mustard [31]. Reprinted from *Executive Summary of Chemical Warfare Agents and their Hydrolysis Products from the US EPA Standardized Analytical Methods and GC-MS Analytical Method for the Analysis of Chemical Warfare Agent Degradation Products* listed in the EPA Standardized Analytical Methods US EPA, Theodore A. Haigh

Figure 4.6 shows the main hydrolysis products of Sulfur Mustard [31].

A selective, direct and relatively rapid method has been developed for the determination of thiodiglycol (TDG) in aqueous samples by using microcolumn liquid chromatography coupled on-line with sulfur flame photometric detection using large-volume injections and peak compression [32]. Figure 4.7 reports the separation improvement with the addition of *n*-propanol to a TDG sample. The combined effect of peak compression by displacement with *n*-propanol

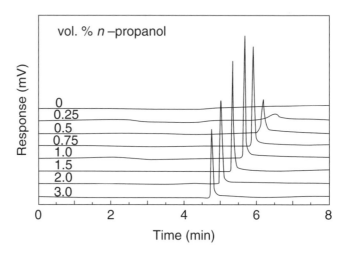

Figure 4.7 Micro-LC–S-FPD chromatograms obtained after the addition of various percentages of *n*-propanol to a TDG sample. Final TDG concentration, $10\,\mu g\,mL^{-1}$ [32]. Reprinted from Determination of the sulfur mustard hydrolysis product thiodiglycol by microcolumn liquid chromatography coupled on-line with sulfur flame photometric detection using large-volume injections and peak compression, Edwin W.J. Hooijschuur, Charles E. Kientz, Udo A. Th. Brinkman *Journal of Chromatography A* 849(2):433–44, 1999, with permission from Elsevier

and large-volume injections (10 mL) resulted in detection limits of $0.25\,mg\,mL^{-1}$ TDG with a total analysis time of 12 min [32].

Low levels of sulfur-containing degradation products of Sulfur Mustard can be identified by hyphenated techniques, such as liquid or gas chromatography coupled to inductively coupled plasma mass spectrometry (ICP-MS). A similar type of chromatography to the one reported above (displacement chromatography), has proved to be very effective for the separation of thiodiglycol and other degradation products of Sulfur Mustard, the BHET-alkenes reported in Figure 4.8: bis(2-hydroxyethyl thio) methane (BHETM), 1,2-bis (2-hydroxyethyl thio) ethane (BHETE), 1,3-bis(2-hydro

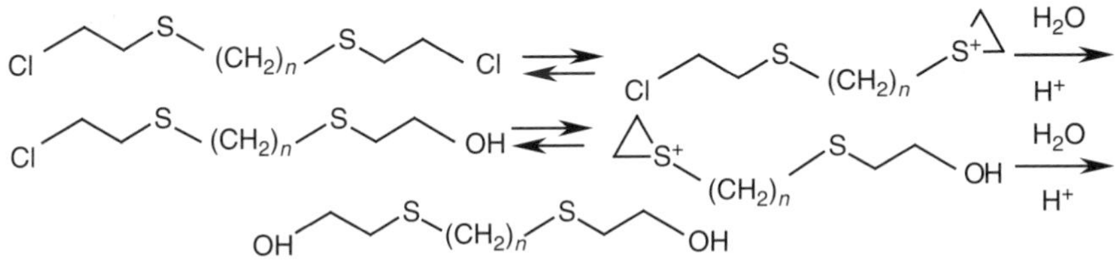

Figure 4.8 Hydrolysis degradation products of Sulfur Mustard

xyethyl thio) propane (BHETPr) and 1,4-bis(2-hydroxyethyl thio) butane (BHETBu) [33].

The described study used displacement chromatography coupled to ICP-MS. The displacers used were 2-methyl-3-pentanol, 3-pentanol and 2,2-dimethyl-3-pentanol. Table 4.2 shows the detection limits based in a $10\,\mu g\,mL^{-1}$ mixture of all five degradation products. The method has also been applied to complex matrices, such as synthetic urine and river water (Little Miami River, Cincinnati, OH, USA) [34].

Thiodiglycol can also be detected by GC-MS. Tomkins and Sega [35] reported a highly-sensitive analytical procedure for determining thiodiglycol in groundwater. Samples are initially fortified with 3,3'-thiodipropanol (surrogate), then both species are extracted using sequential solid phase extraction with both C_{18} and Ambersorb 572 columns.

Table 4.2 Detection limits based in a 10 µg mL^{-1} mixture of all five hydrolysis degradation products of Sulfur Mustard

Parameter	TDG	BHETM	BHETE	BHETPr	BHETBu
Retention time (min)	3.3	7.4	10.2	18.8	29.2
LOD, 3σ (ng mL^{-1})	5 ± 10	35 ± 10	79 ± 10	99 ± 10	73 ± 10
Regression coefficient (r^2)	0.965	1.00	0.998	0.998	0.999

The C_{18} column, which removes extraneous groundwater components, is discarded; the Ambersorb 572 column is dried thoroughly before eluting polar components with a small volume of dichloromethane. The extract is taken to dryness using dry flowing nitrogen, and the resulting residue is derivatized using MTBSTFA [(N-methyl-N-(tert-butyldimethylsilyl)trifluoroacetamide] and pyridine. The derivatized products are diluted to a final volume with toluene, chromatographed using a fused-silica capillary column, and detected with a quadrupole mass spectrometric detector in its selective-ion mode. Two independent statistically unbiased procedures were used to evaluate the detection limits for thiodiglycol; the values ranged between 4 and $16\,\mu g\,L^{-1}$ in groundwater [35].

Nitrogen Mustard gives rise to three alkyl ethanolamines: N-methyldiethanolamine (MDEA), N-ethyldiethanolamine (EDEA) and triethanolamine (TEA), (degradation/precursor products of HN-1, HN-2 and HN-3 blister agents) (Figure 4.9). An electroanalytical method, a poly(dimethylsiloxane) (PDMS) microfluidic device with contactless conductivity detection for the determination of Nitrogen Mustard degradation products, has been reported in the literature by Ding and Rogers [36]. The aforementioned degradation products were analysed by microchip capillary electrophoresis (CE). The original PDMS channel was coated with poly(ethyleneimine) (PEI) to improve the separation of the three ethanolamines. Experimental conditions for the separation and detection processes were optimized to yield well defined separation and high sensitivity. The response times for the three ethanolamines were less than 5 min, the detection limits were $2.0-4.0\,mg\,L^{-1}$ and the relative standard derivations for the migration times and peak heights were 1.6–2.3 and 4.1–5.7%, respectively. Applicability of this method for natural lake and tap water samples was

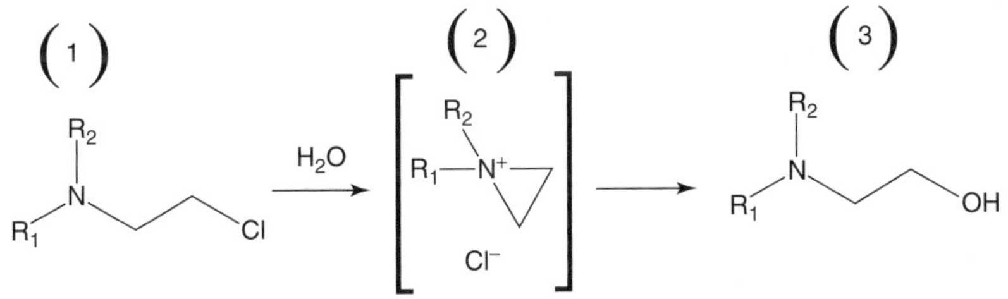

	R$_1$	R$_2$	Nitrogen Mustard
HN1	C$_2$H$_5$	CH$_2$CH$_2$Cl	Steps 2–3 repeated for each CH$_2$CH$_2$Cl group.
HN2	CH$_3$	CH$_2$CH$_2$Cl	
HN3	CH$_2$CH$_2$Cl	CH$_2$CH$_2$Cl	

Figure 4.9 Degradation products of Nitrogen Mustard: *N*-methyldiethanolamine (MDEA), *N*-ethyldiethanolamine (EDEA) and triethanolamine (TEA) [31]. Reprinted from *Executive Summary of Chemical Warfare Agents and their Hydrolysis Products from the US EPA Standardized Analytical Methods and GC-MS Analytical Method for the Analysis of Chemical Warfare Agent Degradation Products* listed in the EPA Standardized Analytical Methods US EPA, Theodore A. Haigh

also demonstrated. Compared with conventional analytical methods, this miniaturized system offers promise for on-site monitoring of degradation products of the Nitrogen Mustard class of chemical warfare agents, with advantages of cost-effective construction, simple operation, portability and the requirement of small sample volumes [36].

Liquid chromatography coupled to mass spectrometry also represents a choice for screening of Nitrogen Mustard degradation products. A fast and effective qualitative screening procedure was developed for these compounds by Chua *et al.* [37], using liquid chromatography–mass spectrometry (LC-MS), which eliminates the need for additional sample handling and derivatization [which is required for gas chromatographic–mass spectrometric (GC-MS) analysis].

A liquid chromatograph with a mixed-mode column and isocratic elution was used for the characterization of the degradation products in two matrices (spiked water and decontamination emulsion). The results demonstrated that N-methyldiethanolamine (MDEA), N-ethyldiethanolamine (EDEA) and triethanolamine (TEA) are not the major degradation products of their respective Nitrogen Mustards, but the partially hydrolysed Nitrogen Mustards are more prominently in evidence [37].

A method for the determination of 2-chlorovinyl arsenous acid (CVAA) and 2-chlorovinyl arsonic acid (CVAOA), which are degradation compounds of the chemical warfare agent Lewisite, was examined by high-performance liquid chromatography–inductively coupled plasma–mass spectrometry (HPLC–ICP-MS) by Kinoshita *et al.* [38]. Inertsil C_8 was suitable as the column and the mobile phase consisted of 0.1% formic acid–acetonitrile (80:20). These compounds were detected with good sensitivity in a short time and separated from inorganic arsenicals and diphenylarsinic acid (DPAA) and phenylarsonic acid (PAA), which are degradation compounds of diphenylchloroarsine and phenyldichloroarsine, respectively. The arsenic detection limits for CVAA and CVAOA were 0.2 and $0.1\,\mathrm{ng\,mL^{-1}}$. In addition, a dynamic reaction cell and oxygen as the reaction gas were applied, and then arsenic was detected as AsO^+ (m/z 91) in order to prevent interference by $ArCl^+$ (m/z 75). This method was applied to the analysis of urine obtained from a CVAA-administered mouse and CVAOA was detected as the main metabolite. Thus, the speciation analysis of arsenic compounds derived from chemical warfare agents was achieved by HPLC–ICP-MS [38].

4.4 ANALYSIS OF STERNUTATOR DEGRADATION PRODUCTS

Until recently, the degradation products of the sternutator agents Clark I and Clark II had been rarely studied. The main degradation products consist of diphenylarsinic acid (DPAA), phenylarsinic acid (PAA), phenylarsinic oxide (PAO), triphenylarsine (TPA) and triphenylarsine oxide (TPAO). All of these compounds are nonvolatile, so the analysis by gas chromatography (GC) (the most popular analysis method for Clark I and II) is somewhat impractical unless derivitization takes place. The analysis calls for long, difficult derivatizations, which can lead to complicated analyses. Because of this, liquid chromatography (LC) coupled to a form of mass spectrometry (MS) has become the premier method used to analyse the sternutator degradation products. Other methods include gas chromatography and zwitterionic hydrophilic interaction chromatography.

Successful derivatization is the key step for GC analysis of nonvolatile analytes. DPAA along with PAA have been derivatized using n-propanethiol under normal and acidic conditions [39, 40]. Thiol derivatization is a popular technique and allows for easy determination of the analyte. In a study done by Hanaoka et al. [39], the efficiency of a thiol derivatization was examined. For the degradation products, acidic conditions were required in order to properly facilitate a reaction. Most analytes had almost 100% recovery rates from the derivatization. The derivatization was also applied to a well water sample collected from Kamisu City. They were able to identify PAA that was present in the water (Figure 4.10). These workers also explored the possibility of analysing TPA by GC-FID without derivatization. Hanaoka et al. were able to use an on-column injection technique to avoid thermal breakdown of the analyte and obtained a good linear range for TPA without derivatization [39].

Liquid chromatography is a very popular analytical method due to its sensitivity, selectivity and the relatively easy sample preparation. It can also be coupled to different mass spectrometry techniques to offer excellent separation and mass spectral identification of a mixture of analytes. LC-UV has been used for the identification of DPAA and PAA. Ishizaki *et al.* used HPLC for the separation of DPAA under reversed phase gradient conditions. They employed the use of an L-Column ODS. Their method helped to determine that there was a high concentration of DPAA in a sample of well water (15 ppm) [40]. Hanaoka and coworkers also analysed DPAA and PAA using HPLC with limits of detection as low as $1\,\mu g\,mL^{-1}$ [39].

Although LC methods have proved to be efficient and simple, matching of the retention times is the only identification that they provide. This means that all the analytes must be fully separated for total mixture identification. However, when LC is coupled to ESI-MS, analytes that are not fully separated can be identified using their mass to charge ratios. Wada and coworkers employed positive-mode MS to identify PAA, DPAA and PAO with LODs of 0.05, 0.0001 and $0.01\,mg\,mL^{-1}$, respectively [41]. Each of the nine analytes were able to be identified in a mixture from a soil sample and with extraction recoveries of >80%. PAO did have a poor extraction recovery due to its solubility with water. As a result of this successful extraction, Wada *et al.* developed a robust method that should be applicable to most aqueous samples.

Zwitterionic hydrophilic interaction chromatography (ZIC-HILIC) has also been used as a method of separating DPAA and PAO [42]. This separation technique relies on a partitioning mechanism between the mobile phase (high

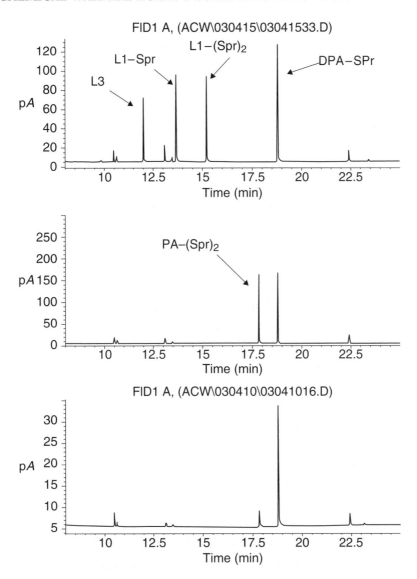

Figure 4.10 Chromatograms from GC-FID analysis of: upper chromatogram, Lewisite and DPAA; middle chromatogram, DPAA and PAA; and lower chromatogram, well water sample from Kamisu City. PAA could be identified in the water sample [39]. S. Hanaoka, *et al.*, Determination of diphenylarsenic compounds related to abandoned chemical warfare agents in environmental samples, *Applied Organometallic Chemistry*, 2005, 19, 265–275, with permission from Wiley

organic content) and a water-rich layer on the column bed. The use of a mainly organic solvent in the separation makes it an attractive method for coupling to ESI-MS. The high organic content of the mobile phase allows for better ionization into the mass spectrometer and increased sensitivity. Xie *et al.* [42] used ZIC-HILIC for the separation of PAA, PAO and other organoarsenicals. They were able to successfully separate and identify nine different analytes using isocratic (78% acetonitrile) and gradient techniques (78–70% acetonitrile), while using both ESI-MS and ICP-MS in parallel for their detection (Figure 4.11).

The coupling of HPLC to ICP-MS offers multielement analysis that is fast, robust and reproducible. ICP-MS also offers excellent sensitivity with detection limits in the parts per trillion range for most elements. The separation and detection of sternutator agent degradation products by HPLC–ICP-MS offers two forms of identification: retention time and elemental identification. It also offers the ability to identify mixtures of degradation products in complex matrices.

Kobayashi and Hirano were able to identify DPAA in urine samples through analysis by HPLC–ICP-MS [43]. They studied the metabolic effect of DPAA in the body by supplementing monkeys with DPAA and then testing excretion amounts of arsenic and arsenic levels in the hair. Total arsenic was determined by ICP-MS in addition to speciation of the arsenic from the samples. Kobayashi and Hirano also used a gel filtration column to obtain two different arsenic peaks, one that was protein bound DPAA and the other free arsenic. The detection limit for total arsenic using ICP-MS was found to be $0.02 \, \text{mg} \, \text{mL}^{-1}$.

DPAA in humans has also been investigated. Kinoshita *et al.* [44] tested different columns and mobile phases for the detection of phenylarsenic compounds in urine from patients, mice and environmental samples of groundwater

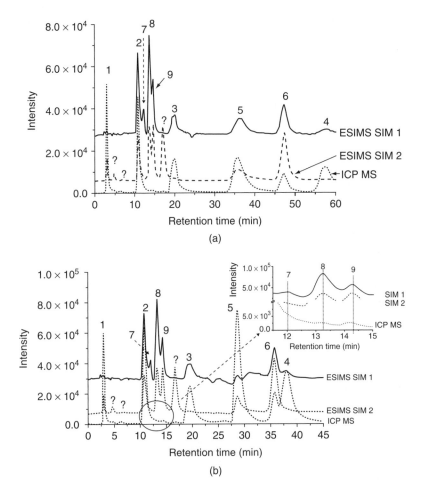

Figure 4.11 Separation of nine different organoarsenicals including PAO (peak 1) and PAA (peak 2). (a) Isocratic separation with 78% acetonitrile; and (b) gradient separation with 78–70% acetonitrile. ICP-MS and ES-IMS were used in parallel for elemental and molecular identification [42]. D. Xie, *et al.*, Separation of organoarsenicals by means of zwitterionic hydrophilic interaction chromatography (ZIC-HILIC) and parallel ICPMS/ESI-MS detection, *Eng. Life Sci.*, 2008, 8, 582–588 with permission from Wiley

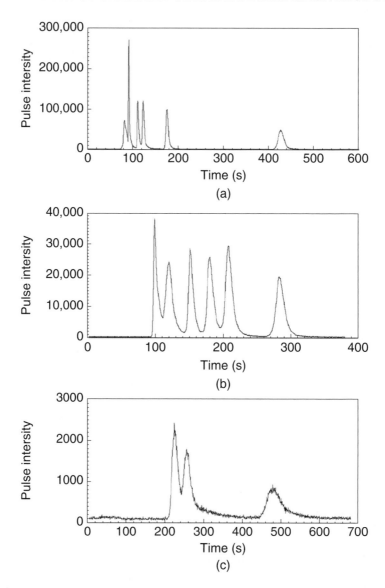

Figure 4.12 Chromatograms of five standard phenylarsenic compounds (PDMAO, PMAA, PAA, DPMAO and DPAA) separated on three different stationary phases. (a) C₄ column, (b) CN column and (c) NH₂ column. The C₄ column gave a more efficient separation and was used in further studies [44]. Kinoshita K. *et al*, Urine analysis of patients exposed to phenylarsenic compunds via accidental pollution, *Journal of Chromatography B*, 867, 2008 with permission from Elsevier

and rice. The columns tested were C_4, cyano and NH_2 columns with their appropriate mobile phases. After optimizing each separation, the C_4 gave the best overall resolution (Figure 4.12) and was used for further testing on the urine samples. When coupled to ICP-MS, these workers used a DRC (Dynamic Reaction Cell) and employed the use of oxygen in the cell to overcome any interferences of m/z 75. Overall, Kinoshita *et al.* were able to develop a robust method with detection limits in the range of $0.1–0.5 \, ng \, mL^{-1}$ for arsenic in the determination of phenyarsenic species in urine and environmental samples.

Owing to the contamination of well water in Kamisu City, Japan, Arao *et al.* employed HPLC–ICP-MS ($ArCl^+$) to study the uptake of arsenic from contaminated soil by rice [45]. In this study, arsenic was extracted from rice grain, straw and soil by a hot-plate digestion with nitric acid. A gradient reversed phase separation on a C_{18} column was used in order to separate the various arsenic species. DPAA, PAA and different methylated forms of PAO were found in the soil samples. Also, methylated PAA was detected in the rice grain and MPAA (methylphenylarsinic acid), DMPAO (dimethylphenylarsine oxide) and MDPAO (methyldiphenylarsine oxide) were detected in the straw. These workers considered that the methylation of the analytes occurs under flooding conditions and these results warrant further investigations as to the location of arsenicals in the shoots and grains of the rice.

REFERENCES

[1] Organization for the Prohibition of Chemical Weapons (2010) *www.OPCW.org* (accessed 25 January 2010).
[2] Munro, N.B., Talmage, S.S., Griffin, G. D., *et al.* (1999) *Environ. Health Perspect.*, **107** (12), 933–974.

[3] John, H., Worek, F., Thiermann, H. (2008) *Anal. Bioanal. Chem.*, **391** (1), 97–116.

[4] Xu, L., Gong, X.Y., Lee, H.K., Hauser, P.C. (2008) *J. Chromatogr. A*, **1205** (1-2), 158–162.

[5] Le Moullec, S., Begos, A., Pichon, V., Bellier, B. (2006) *J. Chromatogr. A*, **1108** (1), 7–13.

[6] Saradhi, U.V.R.V., Prabhakar, S., Reddy, T.J., Vairamani, M. (2006) *J. Chromatogr. A*, **1129** (1), 9–13.

[7] Kanaujia, P.K., Pardasani, D., Gupta, A.K. *et al.* (2007) *J. Chromatogr. A*, **1161** (1-2), 98–104.

[8] Subramaniam, R., Aastot, C., Nilsson, C., Oestin, A. (2009) *J. Chromatogr. A*, **1216** (48), 8452–8459.

[9] Rearden, P., Harrington, P.B. (2005) *Anal. Chim. Acta*, **545** (1), 13–20.

[10] Zygmunt, B., Zaborowska, A., Swiatlowska, J., Namiesnik, J. (2007) *Curr. Org. Chem.*, **11** (3), 241–253.

[11] Lancas, F.M., Queiroz, M.E.C., Grossi, P., Olivares, I.R.B. (2009) *J. Sep. Sci.*, **32** (5-6), 813–824.

[12] Black, R.M., Muir, B. (2003) *J. Chromatogr. A*, **1000** (1-2), 253–281.

[13] Creasy, W.R. (1999) *J. Am. Soc. Mass Spectrom.*, **10** (5), 440–447.

[14] Richardson, D.D., Caruso, J.A. (2007) *Anal. Bioanal. Chem.*, **389** (3), 679–682.

[15] D'Agostino, P.A. (2008) Chemical warfare agents, in *Handbook of Analytical Separations*, 2nd edn, vol. 6 (ed. M.J. Bogusz), Elsevier, Amsterdam, pp. 839–872.

[16] Hooijschuur, E.W.J., Kientz, C.E., Brinkman, U.A.T. (2002) *J. Chromatogr. A*, **982** (2), 177–200.

[17] Soederstroem, M.T., Ketola, R.A. (1994) *Fresenius J. Anal. Chem.*, **350** (3), 162–167.

[18] Ciner, F.L., McCord, C.E., Plunkett, R.W. *et al.* (2007) *J. Chromatogr. B: Anal. Technol. Biomed. Life Sci.*, **846** (1-2), 42–50.

[19] Mawhinney, D.B., Hamelin, E.I., Fraser, R. *et al.* (2007) *J. Chromatogr. B: Anal. Technol. Biomed. Life Sci.*, **852** (1-2), 235–243.

[20] Mawhinney, D.B., Stanelle, R.D., Hamelin, E.I., Kobelski, R.J. (2007) *J. Am. Soc. Mass Spectrom.*, **18** (10), 1821–1826.

[21] Owens, J., Koester, C. (2009) *J. Agric. Food Chem.*, **57** (18), 8227–8235.

[22] Mazumder, A., Gupta, H.K., Garg, P. *et al.* (2009) *J. Chromatogr. A*, **1216** (27), 5228–5234.

[23] Richardson, D.D., Sadi, B.B.M., Caruso, J.A. (2006) *J. Anal. At. Spectrom.*, **21** (4), 396–403.

[24] Kubachka, K.M., Richardson, D.D., Heitkemper, D.T., Caruso, J.A. (2008) *J. Chromatogr. A*, **1202** (2), 124–131.

[25] Richardson, D.D., Caruso, J.A. (2007) *Anal. Bioanal. Chem.*, **388** (4), 809–823.

[26] Kroening, K.K., Richardson, D.D., Kubachka, K.M., Caruso, J.A. (2008) *Spectroscopy*, **23** (9), 34, 36–41.

[27] Kanu, A.B., Haigh, P.E., Hill, H.H. (2005) *Anal. Chim. Acta*, **553** (1-2), 148–159.

[28] Kolakowski, B.M., D'Agostino, P.A., Chenier, C., Mester, Z. (2007) *Anal. Chem.*, **79** (21), 8257–8265.

[29] Lagarrigue, M., Bossee, A., Begos, A. *et al.* (2006) *J. Chromatogr. A*, **1137** (1), 110–118.

[30] Xu, L., Hauser, P.C., Lee, H.K. (2009) *J. Chromatogr. A*, **1216** (31), 5911–5916.

[31] EPA Standardized Analytical Methods US EPA (2007) *Executive Summary of Chemical Warfare Agents and their Hydrolysis Products from the US EPA Standardized Analytical Methods and GC-MS Analytical Method for the Analysis of Chemical Warfare Agent Degradation Products*, http://www.epa.gov/nhsrc/pubs/paperCWAsinSAM092608.pdf (accessed on 23 June, 2010).

[32] Hooijschuur, E.W.J., Kientz, C E., Brinkman, U.A.T. (1999) *J. Chromatogr. A*, **849**, 433–444.

[33] Drasch, G., Kretschmer, E., Kauert, G., von Myer, J. (1987) *Forensic Sci.*, **32**, 1788–1793.

[34] Kroening, K., Richardson, D., Afton, S., Caruso, J. (2009) *Anal. Bioanal. Chem.*, **393**, 1949–1956.

[35] Tomkins, B.A., Sega, G.A. (2001) *J. Chromatogr. A*, **911**, 85–96.

[36] Ding, Y., Rogers, K. (2008) *Electroanalysis*, **20**, 2192–2198.

[37] Chua, H.-C., Lee, H.-S., Sng, M.-T. (2006) *J. Chromatogr. A*, **1102**, 214–223.

[38] Kinoshita, K., Shikino, O., Seto, Y., Kaise, T. (2006) *Appl. Organomet. Chem.*, **20**, 591–596.

[39] Hanaoka, S., Nagasawa, E., Nomura, K. *et al.* (2005) *Appl. Organomet. Chem.*, **19**, 265–275.

[40] Yanaoka, T., Nakamura, M., Hakuta, T. *et al.* (2004) *J. Health Sci.*, **51**, 130–137.
[41] Wada, T., Nagasawa, E., Hanaoka, S. (2006) *Appl. Organomet. Chem.*, **20**, 573–579.
[42] Xie, D., Mattusch, J., Wennrich, R. (2008) *Eng. Life Sci.*, **8**, 582–588.
[43] Kobayashi, Y., Hirano, S. (2008) *Arch. Toxicol.*, **82**, 553–561.
[44] Kinoshita, K., Shikino, O., Seto, Y., Kaise, T. (2008) Urine analysis of patients exposed to phenylarsenic compounds via accidental pollution. *J. Chromatogr. B*, **867** (2), 88–179.
[45] Arao, T., Maejima, Y., Baba, K. (2009) *Environ. Sci. Technol.*, **43**, 1097–1101.

Appendix

AChE: Acetylcholinesterase
ACN: Acetonitrile
Adamsite: 10-chloro-5,10-dihydrophenarsazine (DM)
AED: Atomic Emission Detection
API: Atmospheric pressure ionization
APCI: Atmospheric pressure chemical ionization
ASE: Accelerated solvent extraction
AUES: American University Experiment Station
BHETM: Bis(2-hydroxyethylthio)methane
BHETBu: 1,4-bis(2-hydroxyethylthio)butane
BHETE: 1,2-bis(2-hydroxyethylthio)ethane
BHETPr: 1,3-bis(2-hydroxyethylthio)propane
BZ: 3-Quinuclidinyl benzilate
CDC: Center for Disease Control and Prevention
CE: Capillary electrophoresis
CH-COSY: Carbon-Hydrogen Correlation Spectroscopy
CI: Chemical Ionization
Clark I: diphenylchloroarsine, DA

Analysis of Chemical Warfare Degradation Products, First Edition. Karolin K. Kroening,
Renee N. Easter, Douglas D. Richardson, Stuart A. Willison and Joseph A. Caruso.
© 2011 John Wiley & Sons, Ltd. Published 2011 by John Wiley & Sons, Ltd.

Clark II: diphenylcyanoarsine, DC
CMPA: cyclohexyl methylphosphonic acid
CTAB: cetyltrimethylammoniumbromide
CVAA: chlorovinyl arsonous acid
CVAO: chlorovinyl arsonous oxide
CWA: Chemical Warfare Agent
CWC: Chemical Weapons Convention
CWS: Chemical Warfare Service
CX: phosgene oxime
DA: Diphenylcyanoarsine (Clark I)
DART: Direct analysis in realtime
DC: Diphenylchloroarsine (Clark II)
DEA: Diethanolamine
DESH: Diisopropylaminoethane thiol
DESI: Desorption Electrospray Ionization
DIMP: Diisopropyl methylphosphonic acid
DIPC: Diisopropyl carbodiimide
DM: Adamsite
DMA: Dimethylarsinic acid
DPAA: Diphenylarsinic acid
DSC: Differential Scanning Calorimetry
EA 2192: S-(2-diisopropylaminoethyl)
    methylphosphonothioic acid
EA 4196: bis(diisopropylaminoethyl) disulfide
ECD: Electron Capture Detector
EDEA: N-ethyldiethanolamine
EDS: Energy dispersive X-ray analysis
EHDAP: ethyl hydrogen dimethylamidophosphate sodium
    salt
EI: Electron Impact Ionization
EMPA: ethyl methylphosphonic acid
EPR: electron paramagnetic resonance
EI: Electron Impact
ESI: electrospray ionization

EtPA: 10-ethyl-5,10-dihydrophenarsazine
FAB: Fast atom bombardment
FDA: Food and Drug Administration
FID: Flame Ionization detector
FPD: Flame photometric detector
FTIR: Fourier transform infrared spectroscopy.
GA: Tabun
GB: Sarin
GC: Gas Chromatography
GD: Soman
GF: Cyclosarin
HBDP: 2,2',4,4',6,6'-hexabromodiphenylamine
HCl: Hydrochloric acid
HD: Sulfur mustard
HN1: Bis(2-chloroethyl)ethylamine - nitrogen mustard 1
HN2: Methyl-bis-(2-chloroethyl)amine - nitrogen mustard 2
HN3: Tris(2-chloroethyl)amine - nitrogen mustard 3
HPLC: High Performance Liquid Chromatography
IARC: International Agency for Research on Cancer
IBHMP: isobutyl hydrogen methylphosphonate
ICAM: Improved Chemical Agent Monitor
ICI: Imperial Chemical Industries
ICPMS: Inductively coupled plasma mass spectrometry
IMPA: isopropyl methylphosphonic acid
IMS: Ion Mobility Spectroscopy
INN: dimercaprol or british anti lewisite
IR: Infra-red
LC: Liquid Chromatography
LC_{50}: Median lethal concentration
LD_{50}: Median lethal dose
LC_{LO}: Lowest lethal concentration
LD_{LO}: Lowest lethal dose
LLE: liquid-liquid extraction
LLLME: liquid-liquid-liquid microextraciton

LOA: limit of alarm
LOD: Limit of Detection
LOQ: Limit of Quantitation
LSE: liquid solvent extraction
MPA: Methylphosphonic acid
MDEA: N-methyldiethanolamine
MS: Mass Spectrometry
MS-MS: Tandem Mass Spectrometry
MTBSTFA: N-methyl-N-(tert-butyldimethylsilyl)
  trifluoroacetamide
NMR: Nuclear Magnetic Resonance Spectroscopy
NOESY: Nuclear Overhauser Enhancement Spectroscopy
NPD: Nitrogen-Phosphorus Detector
OPCW: Organization for the Prohibition of Chemical
  Weapons
PAA: Phenylarsinic acid
2-PAMCl: 2-Parlidoxime chloride
PAO: phenylarsinic oxide
PBI: Particle Beam Ionization
PDMS: Poly(dimethylsiloxane)
PEI: Poly(ethyleneimine)
PFIB: Perfluoroisobutylene
PFPD: pulsed flame photometric detector
PGNAA: Prompt gamma-ray neutron activation analysis
PMPA: pinacolyl methylphosphonic acid
POM: Polyoxometalates
PPT: Part per trillion
PRPS: Powered Respirator Protective Suit
PTV: programmed temperature vaporizing
RD_{50}: 50% Respiratory Rate Decrease
RVX: see VR
SAW: Surface Acoustic Wave
SBSE: stir bar sorptive extraction
SCD: Sulfur Chemiluminescence Detector

SEM: Scanning Electron Microscopy
SIM: Single or Selected Ion Monitoring
SPE: Solid Phase Extraction
SPME: Solid Phase Microextraction
TBA: Tributylamine
TBDMS: Tert-butyldimethylsilyl
TD-IMS: Thermal Desorption-Ion Mobility Spectroscopy
TDG: Thiodiglycol
TEA: Triethanolamine
TEP: Triethyl phosphate
TG: Thermogravametric
TMS: Trimethylsilyl
TPA: Triphenylarsine
TPAO: Triphenylarsine oxide
TS: Thermospray
UK: United Kingdom
UV: Ultra-violet
VE: O-ethyl-S-[2-(diethylamino)ethyl]
  ethylphosphonothioate
VG: O, O-diethyl-S-[2-diethylaminoethyl]
  phosphorothioate (Amiton)
VM: O-ethyl-S-[2-(diethylamino)ethyl]
  methylphosphonothioate
VR or RVX: N,N-diethyl-2-(methyl-(2-methylpropoxy)
  phosphoryl)sulfanylethanamine-(Russian VX)
VX: O-ethyl S-[2-(diisopropylamino)ethyl]
  methylphosphonothioate
WWI: World War I
WWII: World War II
ZIC-HILIC: Zwitterionic hydrophilic interaction
  chromatography

Index

Analysis of Chemical Warfare Degradation Products, First Edition. Karolin K. Kroening,
Renee N. Easter, Douglas D. Richardson, Stuart A. Willison and Joseph A. Caruso.
© 2011 John Wiley & Sons, Ltd. Published 2011 by John Wiley & Sons, Ltd.